# Electrical and Magnetic Methods of Non-destructive Testing

# NON-DESTRUCTIVE EVALUATION SERIES

Non-destructive evaluation now has a central place in modern technology both as a means of evaluating materials and products as they are manufactured and for confirmation of fitness for purpose while they are in use.

This series provides in-depth coverage of the wide range of techniques that are now available for the non-destructive evaluation of materials. Each volume will contain material that is relevant to final year undergraduates in Engineering, Materials Science and Physics in addition to post graduate students, experienced research workers, and practising engineers. In some cases they will be written with taught courses in mind, while other texts will be for the qualified engineer or scientist who wishes to become familiar with a new topic at research level.

**Series editors**
*Professor S. Palmer*　　*Professor W. Lord*
*Department of Physics*　*Department of Electrical and Computer Engineering*
*University of Warwick*　*Iowa State University*
*Coventry*　　　　　　　*Iowa*
*UK*　　　　　　　　　　*USA*

**Titles available**
Numerical Modeling for Electromagnetic Non-Destructive Evaluation
Nathan Ida

Industrial Radiology
R. Halmshaw

Ultrasonic Methods of Non-destructive Testing
Blitz and Simpson

# Electrical and Magnetic Methods of Non-destructive Testing

*Second edition*

### Jack Blitz

*Formerly Senior Lecturer
Department of Physics
Brunel University*

**CHAPMAN & HALL**
London · Weinheim · New York · Tokyo · Melbourne · Madras

Published by Chapman & Hall, 2–6 Boundary Row, London SE1 8HN

Chapman & Hall, 2–6 Boundary Row, London SE1 8HN, UK

Chapman & Hall GmbH, Pappelallee 3, 69469 Weinheim, Germany

Chapman & Hall USA, 115 Fifth Avenue, New York, NY 10003, USA

Chapman & Hall Japan, ITP-Japan, Kyowa Building, 3F, 2-2-1 Hirakawacho, Chiyoda-ku, Tokyo 102, Japan

Chapman & Hall Australia, 102 Dodds Street, South Melbourne, Victoria 3205, Australia

Chapman & Hall India, R. Seshadri, 32 Second Main Road, CIT East, Madras 600 035, India

First edition 1991
© IOP Publishing
Second edition 1997
© 1997 Chapman & Hall

Typeset in Palatino by AFS Image Setters Ltd, Glasgow
Printed in Great Britain by TJ International Ltd, Padstow, Cornwall

ISBN 0 412 79150 1

Apart from any fair dealing for the purposes of research or private study, or criticism or review, as permitted under the UK Copyright Designs and Patents Act, 1988, this publication may not be reproduced, stored, or transmitted, in any form or by any means, without the prior permission in writing of the publishers, or in the case of reprographic reproduction only in accordance with the terms of the licences issued by the Copyright Licensing Agency in the UK, or in accordance with the terms of licences issued by the appropriate Reproduction Rights Organization outside the UK. Enquiries concerning reproduction outside the terms stated here should be sent to the publishers at the London address printed on this page.
  The publisher makes no representation, express or implied, with regard to the accuracy of the information contained in this book and cannot accept any legal responsibility or liability for any errors or omissions that may be made.

A catalogue record for this book is available from the British Library

Library of Congress Catalog Card number: 97-69699

∞ Printed on permanent acid-free text paper, manufactured in accordance with ANSI/NISO Z39.48-1992 and ANSI/NISO Z39.48-1984 (Permanence of Paper).

# Contents

**Preface to the first edition**     xi

**Preface**     xi

**1 Introduction**     1

    1.1 General considerations     1
    1.2 Methods of non-destructive testing     4
    1.3 Electrical and magnetic methods     4
    1.4 Choosing a method     5
    1.5 Automation in non-destructive testing     9
    1.6 Conclusion     10
    1.7 Further reading     11

**2 Fundamental theory**     13

    2.1 General considerations     13
    2.2 Electrical conductivity and resistivity     13
    2.3 Dielectric materials     15
    2.4 Electromagnetism     19
    2.5 Alternating currents     23
    2.6 Circuit networks     31
    2.7 Ferromagnetic materials     34
    2.8 Electromagnetic radiation     39

**3 Magnetic methods**     43

    3.1 Introduction     43
    3.2 Flux leakage methods     44
    3.3 Magnetic particle inspection (MPI)     48
    3.4 Magnetic tape inspection (magnetography)     65
    3.5 Quantitative flux leakage detectors     67
    3.6 Quantitative flux leakage applications     74
    3.7 Magnetization and hysteresis methods     82
    3.8 Less common magnetic techniques     89

## 4 Eddy current principles — 94

4.1 Introduction — 94
4.2 Coils encircling defect-free metal rods — 96
4.3 Coils encircling defect-free metal tubes — 104
4.4 Internal coaxial coils in defect-free metal tubes — 108
4.5 Coils scanning the surfaces of defect-free conductors — 109
4.6 Defect modelling — 121

## 5 Eddy current methods — 132

5.1 General considerations — 132
5.2 Fundamental measurements — 132
5.3 Probe design — 137
5.4 Requirements for eddy current measurements — 144
5.5 Basic eddy current tests: measurements — 149
5.6 Basic eddy current defect detection and sizing — 156

## 6 More advanced eddy current testing methods — 162

6.1 Automatic testing — 162
6.2 Multifrequency testing — 163
6.3 Remote field testing — 167
6.4 Lift-off flaw detection — 171
6.5 Pulsed eddy current testing — 173
6.6 Microwave eddy current testing — 176
6.7 Fibre-reinforced plastics — 177
6.8 Neural networks — 179
6.9 Defect imaging — 181

## 7 Microwave methods — 184

7.1 Introduction — 184
7.2 Microwave radiation — 186
7.3 Microwave instrumentation — 195
7.4 Microwave measurements — 202

## 8 Miscellaneous methods — 214

8.1 General considerations — 214
8.2 Potential drop methods — 214
8.3 Resistance strain gauges — 223
8.4 Electrified particle testing — 225
8.5 Direct measurement of resistance or capacitance — 227

## Appendices — 232

A Notes on units — 232
B Standards — 234
C Bessel functions — 237

## Contents

| | | |
|---|---|---|
| D | BASIC programs for predicting impedance components | 241 |
| | D.1 Coils encircling electrically conducting cylindrical rods | 241 |
| | D.2 Coils encircling electrically conducting cylindrical tubes | 243 |
| | D.3 Air-cored coils: scanning the surfaces of electrical conductors | 248 |
| **References** | | **251** |
| **Index** | | **257** |

# Preface to the first edition

This book is intended to help satisfy an urgent requirement for up-to-date comprehensive texts at graduate and senior undergraduate levels on the subjects in non-destructive testing (NDT). The subject matter here is confined to electrical and magnetic methods, with emphasis on the widely used eddy current and magnetic flux leakage methods (including particle inspection), but proper attention is paid to other techniques, such as microwave and AC field applications, which are rapidly growing in importance.

Theoretical analyses relating to the various methods are discussed and the depths of presentation are often governed by whether or not the information is readily available elsewhere. Thus, for example, a considerable amount of space is devoted to eddy current theory at what the author considers to be a reasonable standard and not, as usually experienced, in either a too elementary manner or at a level appreciated only by a postgraduate theoretical physicist.

The inclusion of the introductory chapter is intended to acquaint the reader with some of the philosophy of NDT and to compare, briefly, the relative performances of the more important methods of testing. Chapter 2 provides a summary of the basic electrical and magnetic theory relating to the subject. It is mainly intended for reference rather than for tutorial purposes, although some of the more important aspects of the theory are treated at greater length. The remainder of the chapters discuss the various techniques and applications. An appendix contains information on SI units, British and American Standards relevant to electrical and magnetic testing, a brief introduction to the Bessel and other functions appertaining to eddy current theory and some programs in BASIC for calculating, with a personal computer, the impedance components of eddy current coils used for testing defect-free metal samples.

In the preparation of this work, the author is especially indebted to Professor Dr F. Förster and his colleagues at the Institut Dr Förster and their former UK agents Wells–Krautkramer Ltd for the valuable help and cooperation provided. He is also grateful to Atomic Energy of

Canada Ltd, the British Gas Corporation, CNS Electronics Ltd, Eddy Current Technology Inc., Elcometer Instruments Ltd, Rolls-Royce plc, Technical Software Consultants Ltd, the NDE Centre at University College, London, and to Dr J. Beynon, Dr R. Collins, Professor W. D. Dover, Mr W. G. King, Mr M. Hajian, Professor C. A. Hogarth, Professor W. Lord, Dr L. Morgan, Dr S. R. Oaten, Mr D. G. Rogers, Commander G. M. B. Selous and his colleagues, Dr E. L. Short and Mr D. Topp for their kind help in different ways. Thanks are also given to the various sources of information acknowledged in the text. The author is grateful to Professor D. C. Imrie, dean of the Faculty of Science at Brunel University, for allowing him the use of the facilities of the Department of Physics.

**Jack Blitz**
Department of Physics, Brunel University, Uxbridge, UK
March 1991

# Preface

Important new developments in the fields of electrical and magnetic testing have been taken into account for this second edition and the book is slightly longer. There has also been an opportunity to make some minor amendments to the text; I am grateful to Dr Thomas Rose of the University of Cologne for his efforts in this direction.

I am also grateful to Professor Cyril Hogarth, Mr Karol Schlachter and Mrs Kay Lawrence at Brunel University, Mr John Bailes at CNS Electronics Ltd, Mr Michael Borden at Technical Software Consultants Ltd, Mr Pete Burrows at Staveley NDT Technologies and Mr Monty O'Connor, president of Eddy Current Technology Incorporated for their kind help. Thanks are given to the various sources of information acknowledged in the text and to Professor Derek Imrie, pro vice chancellor and head of physics at Brunel University, for allowing the use of the facilities in the department.

**Jack Blitz,**
Wembley, UK.

CHAPTER 1

# Introduction

## 1.1  GENERAL CONSIDERATIONS

Although this book deals specifically with electrical and magnetic methods of testing materials, the subject-matter covered forms part of the wider field of non-destructive testing (NDT), briefly discussed in this chapter. It is worthwhile to consider the general scope of NDT before examining specific methods.

Materials and manufactured products are often tested before delivery to the user to ensure they will meet expectations and remain reliable during a specified period of service. It is essential that any test made on a product intended for future use in no way impairs its properties and performance. Any technique used to test under these conditions is called a non-destructive testing method. A manufacturer often subjects his or her products to NDT to maintain a reputation for quality and reliability. The decision whether or not to perform a test is often governed by economic considerations, such as relating the cost of NDT to that of replacing a component which has failed while in service and taking into account any inconvenience and expense to the user caused by the breakdown. As a compromise, the manufacturer may take a calculated risk and perform tests on a fraction of a given batch of products then apply a statistical analysis to determine whether or not to pass the entire batch as being fit for service; this type of operation is often called quality control. However, the onus of the reliability and safety of a product may finally rest with the user, who should ensure that all the necessary testing has been performed by a properly qualified person.

In some circumstances 100% testing is obligatory, not only at the manufacturing stage but also at specified intervals during future service as with the examinations of critical regions in structures and components used in aircraft and in nuclear energy, chemical processing and offshore fuel installations. These structures and components may be subjected to phenomena such as heavy stressing, fatigue and corrosion. As a result, defects such as cracks are likely to form and grow, often rapidly, to such an extent they cause failure with perhaps disastrous consequences.

Non-destructive testing has gained importance due to the rapid technological progress made during the past half-century in areas such as aviation and nuclear energy, where the risks are high and strict precautions are required. In the past there was a tendency to overdesign, often by specifying excessive amounts of material and the use of redundant supports, so that any overstressing of structures would have been very unlikely. Tests, if any, were often rudimentary and confined mainly to visual and audible inspections; audible inspection could mean striking a casting or forging with an iron bar to produce a sound that depends on whether or not it contains cracks and internals voids. This test is still common but sophisticated equipment is now employed to receive and process the sound. Present-day needs for improvements in the performance of machinery have given rise to the development of new materials and the redesign of structures made from existing materials, with a view to increasing strength and reducing weight. The need to improve dimensional tolerances, perhaps to achieve greater engine efficiencies, is also very important. The factors mentioned earlier have led to widespread applications of non-destructive testing to ensure that safety limits are not exceeded.

Reasons for conducting non-destructive tests:

- To ensure freedom from defects likely to cause failure.
- To ascertain the dimensions of a component or structure.
- To determine the physical and structural properties of any materials in a product.

NDT may be performed after one or more stages of construction, e.g. metal casting, forging, welding and machining, and also during operation with the aim of avoiding failure caused by phenomena such as crack growth and corrosion. For an account of the origin and nature of defects in metal objects at different stages of manufacture and future use, the reader is referred to Taylor (1988).

The need for NDT must be very carefully considered and, for major critical installations, preliminary investigations are necessary to predict any likelihood of the occurrence of structural changes and the appearance of defects which may lead to possible failure. These investigations should be conducted at the design stage of an installation or component and they might well involve both theoretical and experimental work extending over a long period. A properly conducted design should ensure that an object facilitates the use of each method of testing needed to provide complete coverage. An important function of the designer is to study the various stresses to which the components in service may be subjected and the consequent formation and subsequent growth of defects, i.e. use is made of fracture mechanics (Lawn and Wilshaw, 1975). At the same time, consideration must be given to the environments in which the components of the installation might operate during service and also to the possibilities of exposure to very high or very low temperatures, extremes of pressure such as those resulting from very high winds and rough seas, and the presence of corrosive or toxic substances. In many instances, weight may be an important factor and lightness can be achieved by the employment of suitable metal alloys or plastics and by the reduction of thickness. Thickness reduction must be considered in relation to the strength and fracture toughness of the materials used.

The monitoring of any structural changes and defect initiation and growth during these investigations is often called non-destructive evaluation (NDE), for which NDT methods are often employed. An important aid here is the use of computer modelling techniques, usually employing finite element analysis, to predict relationships between defect size and the signal indicated by an appropriate detector. These techniques are of great value in the design of NDT instrumentation and the choice of its optimum operating conditions. An example from eddy current testing (section 4.6) is the evaluation of the size of a probe coil and its optimum frequency of operation for a particular kind of test. Some components and structures should be subjected to continuous NDE monitoring of critical areas during operation; this is known as condition monitoring.

These preliminary investigations lead to a specification for testing the entire structure. It should indicate the precise nature of the testing equipment to be used and the operating parameters, the method of calibrating the equipment and the design of any reference samples, the locations and nature of any defects which may be tolerated (e.g. minimum crack sizes in less critical regions) and the technical qualifications of the persons employed as testers. Any necessary computer hardware and software should be clearly specified.

For more routine operations, including the examination of welds in standard types of pressure vessels, testing specifications such as those given by the British Standards Institution (BSI) and the American Society for Testing and Materials (ASTM) may be readily available (Appendix B).

Finally, there is no point whatsoever in overinspecting regions not crucial to the installations. One should bear in mind how the cost of testing a particular item could well exceed its cash value. The failure of many components located in non-critical areas may provide only minor inconvenience, especially if inexpensive spare parts are readily available and can easily be fitted.

## 1.2 METHODS OF NON-DESTRUCTIVE TESTING

NDT can be performed on metals and non-metals and the method of testing used depends on factors such as the type of material and its dimensions, the environment, the positions of interest within the structure or component under examination, e.g. whether internal or surface defects are sought, and the suitability for data acquisition and processing. Detailed information of the various NDT methods and applications can be found in the further reading at the end of this chapter. A rough classification is as follows:

- Radiological methods: X-rays, gamma rays and neutron beams
- Acoustical and vibrational methods: ultrasonic and mechanical impedance measurements
- Electrical and magnetic methods: eddy current, magnetic flux leakage (including magnetic particle inspection) and microwave testing
- Visual and optical methods: interferometry, holography and dye penetrants
- Thermal methods: infrared radiation and thermal paints

A combination of two or more methods is generally required for the complete inspection of an object, as indicated in section 1.4.

The methods most commonly used are ultrasonic testing, X-radiography, eddy current testing, magnetic particle inspection and dye penetrant application. These methods receive the greatest amount of attention from national and international standards organizations; they attract regular training courses in practical applications and certificates of proficiency in them are awarded by recognized bodies to proven skilled operators.

The methods of testing the object under examination may already be specified but, when a choice of technique is permitted, the testing should be carefully planned with regard to safety, economics and efficiency. Whatever methods are used, even when prespecified, the test object should first be thoroughly inspected as far as possible by eye, perhaps with the aid of a magnifying glass, and by touch. Many cases have occurred in which the use of valuable equipment and time have been wasted in locating flaws which could easily have been seen with the unaided eye in the first instance.

## 1.3 ELECTRICAL AND MAGNETIC METHODS

The electrical and magnetic properties of materials comprise electrical conductivity, electrical permittivity and magnetic permeability, as defined in sections 2.2, 2.3 and 2.4, and are related to the structural and mechanical properties of the materials. Thus grain structure in polycrystalline metals, mechanical strength, hardness and the presence of defects and impurities can be assessed from the results of electrical and magnetic NDT methods.

The principal electrical and magnetic NDT methods are as follows:

- Eddy current testing (Chapters 4, 5 and 6)
- Magnetic particle inspection (Chapter 3)
- Magnetic flux measurements (Chapter 3)
- Electromagnetic microwave testing (Chapter 7)
- Potential drop and AC field techniques (Chapter 8)

Eddy current and magnetic particle methods are most commonly used for electrical and magnetic testing; for obvious reasons, magnetic particle detection is often regarded as being a visual method. The use of eddy current methods is normally restricted to the testing of good electrical conductors, although these methods can be applied to the measurement of the thicknesses of poorly conducting coatings on electrically conducting substrates. Magnetic particle and magnetic flux methods are normally used only for testing ferromagnetic materials. Electromagnetic microwave testing is usually restricted to examining dielectric materials but potential drop methods can be used for testing semiconductors as well as good electrical conductors. Less common electrical methods include electrified particle testing and capacitance measurements (Chapter 8), appropriate only for dielectric materials.

## 1.4 CHOOSING A METHOD

### 1.4.1 General considerations

Four of the more important types of non-destructive investigations are assessment of composition and structure of materials, surface and subsurface defect detection, internal defect detection and measurement of dimensions. Two or more methods may often be considered as being suitable for any one of these tests but this does not necessarily imply they may be regarded as alternative techniques. It often happens that one of the methods is used to complement another or perhaps to verify the findings of the other. A common example of the first instance is to locate any cracks occurring in a steel surface by magnetic particle inspection then to use either the eddy current or AC field method to measure their depths.

### 1.4.2 Material composition and structure

The various factors determining the composition of a material include the nature of the material itself, the constituents of an alloy and the

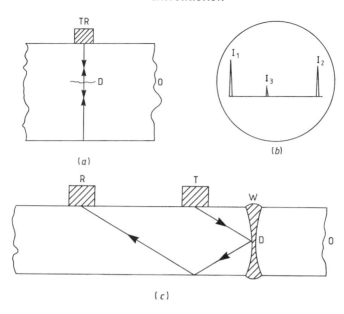

**Figure 1.1** (a) Arrangement for ultrasonic pulse–echo testing with compression waves. (b) Cathode-ray oscilloscope indications. (c) Arrangement for shear wave testing of welded metal. TR = transmitting/receiving probe, T = transmitting probe, R = receiving probe, O = test object, D = defect, W = butt weld. $I_1$, $I_2$ and $I_3$ represent indications for transmission, back echo and defect echo, respectively.

concentrations of the components of a mixture. The factors related to structure include hardness of metals, case hardening depth in alloys, grain size, impurity content and the degree of internal stresses. These factors affect, to different extents, the values of elastic constants, electrical conductivity, magnetic properties and electrical permittivity, each of which varies with temperature.

Elastic constants are related to the speed $c$ of sound and, for an isotropic material of density $\rho$, $c = (q/\rho)^{1/2}$, where $q$ is the appropriate elastic modulus (e.g. either axial or shear modulus for isotropic materials). Strictly speaking, $q$ represents the adiabatic modulus but, for most solids, the adiabatic and isothermal (i.e. static) moduli are very nearly equal to one another. The value of $c$ is best determined by using the ultrasonic pulse–echo method which measures the time of flight of short pulses of high-frequency waves (typically between 1 and 10 MHz) which are reflected from the surface opposite and parallel to the one in contact with the source (Figure 1.1a and b). Ultrasonic waves are generally easy to propagate in metals but not always so in polymers, which often have high acoustic attenuation coefficients.

Electrical permittivities can be determined by using electromagnetic microwaves. These waves are unable to penetrate metals because of the skin effect (section 2.8.2) but they have little difficulty in passing through most polymers. The properties of ferromagnetic metals and alloys can be determined using magnetic methods, e.g. by measuring hysteresis parameters such as coercivity and retentivity (section 2.7). However, if interest is confined only to the subsurface properties of metals, the eddy current method is highly effective for electrical conductivity measurements.

### 1.4.3 Dimension measurements

Dimensions, e.g. thicknesses, are normally determined by visual (i.e. optical), ultrasonic, eddy current and microwave techniques. Visual methods are used where feasible, but when only one boundary of an opaque object is accessible, recourse should be made to one of the other techniques, e.g. the ultrasonic pulse method, which measures the time of flight in a material in which the speed of sound is known. The ultrasonic pulse technique can be used to determine thicknesses of several metres in many metals but its range is generally more limited for non-metals. On the other hand, electromagnetic microwaves can be propagated over comparatively large distances through many poor electrical conductors and have proved to be highly suitable for measuring the dimensions of many plastic objects. When testing non-metals at high temperatures, microwaves have an important advantage over ultrasound: coupling between the probes and the material under examination is possible with microwaves but not with ultrasound.

Microwave testing can also determine thicknesses of parallel-sided metal objects. The eddy current method can be used for measuring the thicknesses of thin metal sheets and foils and also of dielectric coatings, e.g. paints and cladding, on metal substrates.

### 1.4.4 Internal defect detection

Perhaps the most common application of non-destructive testing is the detection of internal flaws. Radiological methods (X-rays and gamma rays) are most widely used for this purpose (Figure 1.2). They have the advantage of producing a readily visible and often highly resolved visual image of the interior of the object under test. The main disadvantage is the need to guard against ionizing radiation hazards; this requires a protected area, e.g. a lead-lined chamber, and the continuous monitoring of the health of personnel. Other drawbacks are the need for film processing facilities and the fact that the depth of penetration of the radiation in the test object may be highly limited, e.g. 50 mm or less in many metals and even less in most polymers, although increased penetration may be achieved with the use of gamma-ray isotopes. Radiological methods have a further disadvantage, their inability to measure depths of defects directly, but this can be overcome by taking two or more exposures in different directions, preferably perpendicular to one another.

The ultrasonic pulse–echo technique has the advantages of being able to locate defect positions in a single operation and of being free from radiation hazards. Either compression (longitudinal) or shear (transverse) waves can be employed. Compression waves are normally used to detect defects lying more or less parallel to the surface of the object under test (Figure 1.1a and b) and shear waves, which can be projected obliquely, for defects oriented roughly perpendicular to the surface, e.g. defects in metal butt welds (Figure 1.1c). Microwaves can also be used to size internal defects in non-metals but generally with a lower degree of precision compared with ultrasound. Further comparisons between ultrasonic and microwave methods are given in section 1.4.3.

### 1.4.5 Surface and subsurface crack detection

The detection of surface-breaking cracks presents few problems and the dye penetrant and magnetic particle methods prove to be highly effective for this purpose. The surface of the test object is covered with very fine particles, which are attracted to any cracks present, and magnified images appear under appropriate illumination. It is important to clean the surface properly before performing the test. The dye penetrant method can be used for testing any solid but it cannot detect subsurface flaws. The magnetic particle technique is suitable only for ferromagnetic metals and requires the application of a strong magnetic field, but it provides a more sensitive and even more certain means of

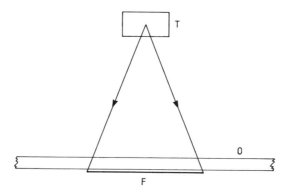

**Figure 1.2** Basic arrangement for X-ray testing: arrows indicate the direction of the beam. T = X-ray tube, O = test object, F = film.

detection and it can also locate subsurface defects. Surface and subsurface defects can also be detected by ultrasonic surface and Lamb (i.e. plate) waves.

An indication of a surface defect by one of these methods presents a choice of either outright rejection of the tested object, if the defect is clearly serious, or testing by another method to ascertain its size. A very small defect can be ground away or even ignored if it is clearly harmless.

More reliable but more complicated ways to measure the depths of cracks and other defects in metals, both surface-breaking and subsurface, include ultrasonic and eddy current testing, quantitative magnetic flux leakage techniques and the AC potential drop and field methods. The ultrasonic technique is also usable for testing non-metals. The microwave method can be applied to detect defects in non-metals and it can also be used for finding and assessing microcracks in metal surfaces.

Where feasible, the simplest ultrasonic method of measuring the depth of a surface-breaking crack is to locate the probe on the surface opposite to the surface containing the crack then to note the time delay of the reflected pulse from its tip. Otherwise, the crack-tip diffraction method can be used, where two probes, one sending the waves at an angle and the other receiving waves scattered from the crack tip, are located on the same surface (Silk, 1977). The depth of a subsurface crack or lamination parallel to the surface can be assessed by locating a flat frequency-response ultrasonic compression wave probe above the crack and using continuous waves at a swept frequency to observe half-wavelength resonance.

The electromagnetic methods are generally more accurate for measuring depths of surface cracks in metal samples. The eddy current technique is more precise for measurements of shallower cracks and AC field measurements for deeper cracks, where the eddy current method becomes less reliable. The magnetic flux leakage method, using Hall effect probes, may also be used for sizing surface-breaking and subsurface cracks, but only for ferromagnetic materials, and it requires the application of a strong magnetic field. It is often used for testing large pipes.

## 1.5 AUTOMATION IN NON-DESTRUCTIVE TESTING

Manual operation of non-destructive testing methods has the advantage that the operator is in a position to make immediate decisions when proceeding from one stage of a test to the next. A well-qualified operator should possess much experience of the idiosyncrasies of the equipment he or she uses for different kinds of testing. Over a period of time, a 'friendly relationship' builds up between the operator and his or her equipment, which may not be possible with automatic testing. However, there is always the possibility of human errors, especially if the operator is distracted or emotionally upset. Recent improvements in the design of NDT equipment and scanning devices, together with the incorporation of microprocessors and computers into testing systems, have made testing extremely reliable. Advances in modelling techniques and in the field of fracture mechanics have enabled an entirely automated testing procedure, eliminating the risks of human error and considerably increasing the speed of testing. Any fault arising in the equipment during its operation can either be corrected automatically by the computer or, if this is not possible, the computer can simply identify it. The detection of a defect in a component, e.g. a crack having more than a prespecified dimension, would allow the computer to divert the component off an assembly line into a rejection bin. Feedback to the most recent stage of manufacture is sometimes used to provide automatic correction of a faulty process.

A brief account of the fundamental application of computers to nondestructive testing has been given by Schroeder (1984).

Automatic testing can be carried out over long distances, e.g. several tens of kilometres. One example is the ultrasonic testing of railway lines, with the equipment located in a car moving over the track and the probes maintained in acoustic contact with the rail. Another is the use of an 'intelligent pig' (section 3.6.2) for testing the walls of pipelines carrying fuel as a liquid or a gas. It consists of a carriage propelled by the pressure of the fluid flow and contains the necessary sensing devices, instrumentation, data processors and power supplies.

## 1.6 CONCLUSION

Electrical and magnetic methods are well represented among the principal NDT techniques, although they take third place to radiological and ultrasonic techniques with regard to amount of use. To make a proper choice of an NDT technique requires an adequate knowledge of the principles behind each method and the types of object for which each method is suitable. This book considers the principles and applications of electrical and magnetic NDT methods; the further reading at the end of this chapter gives more detail on electrical and magnetic methods as well as providing background on more general aspects of NDT and materials testing.

# 1.7 FURTHER READING

There is a scarcity of textbooks on non-destructive testing at the level of this volume. Unfortunately, most books on the subject are currently pitched either at a comparatively low level suitable for operators or at a level more appropriate to specialists and experienced researchers. The more advanced works usually consist of articles by individual authors on diverse topics in narrow fields of NDT or they are reports of conference proceedings. Nevertheless, the books listed below may prove useful. The absence of any particular title may not necessarily imply either unsuitability or lack of quality.

The reader who is interested in the latest developments is recommended to consult the publications listed in the final section. Out-of-print titles may be available for consultation in some major libraries. Materials science books and journals are currently held at the Science Reference and Information Service (SRIS) off Chancery Lane in London. They will move to St Pancras when the new British Library is opened.

## General

Hall B and John V (1988) *Nondestructive Testing* (Berlin: Springer)
Halmshaw R (1987) *Nondestructive Testing* (London: Arnold)
McGonnagle W J (1961) *Nondestructive Testing* (New York: Gordon and Breach)
McMaster R C (ed) (1963) *Nondestructive Testing Handbook* (2 vols) (New York: Ronald) (out of print)
NASA (1973) *Nondestructive Testing – A Survey* (NASA-SP-5113) (Washington DC: NASA) (out of print)

A second edition of the *Nondestructive Testing Handbook*, published by the American Society of Nondestructive Testing (ASNT) at Columbus, Ohio (founder editor, the late Professor R.C. McMaster) is now available and the following volumes have appeared so far:

1. Leak Testing
2. Liquid Penetrant Testing
3. Radiographic Testing
4. Electromagnetic Testing
5. Acoustic Emission Testing
6. Magnetic Particle Testing
7. Ultrasonic Testing
8. Visual and Optical Testing
9. Special Non-destructive Testing Methods
10. Non-destructive Testing Methods Overview

## Electrical and magnetic methods

Bahr A J (1983) *Microwave Nondestructive Testing Methods* (New York: Gordon and Breach)
Betz C E (1967) *Principles of Magnetic Particle Testing* (Chicago: Magnaflux Corporation)

Lord W (ed) (1985) *Electromagnetic Methods of Nondestructive Testing* (New York: Gordon and Breach)
McMaster R C (1986) *Nondestructive Testing Handbook*, 2nd edn, vol 4, *Electromagnetic Testing* (Columbus: ASNT)
McMaster R C (1989) *Nondestructive Testing Handbook*, 2nd edn, vol 6, *Magnetic Particle Testing* (Columbus: ASNT)

## Other methods

Blitz J and Simpson G (1996) *Ultrasonic Methods of Non-destructive Testing* (London: Chapman & Hall)
Halmshaw R (1971) *Industrial Radiology Techniques* (London: Wykeham)
Halmshaw R (1982) *Industrial Radiology – Theory and Practice* (London: Applied Science)
Krautkrämer J and H (1983) *Ultrasonic Testing*, 3rd edn (Berlin: Springer)

## Current developments

### *Journals*

*Insight* (monthly) formerly the *British Journal of Non-Destructive Testing* (Northampton: British Institute of Non-Destructive Testing)
*Journal of Nondestructive Testing Evaluation* (New York: Plenum)
*Materials Evaluation* (monthly) (Columbus: ASNT)
*NDT and E International* (bimonthly) formerly *NDT International* (Oxford: Elsevier)
*Nondestructive Testing and Evaluation* (6 issues per volume) (London: Gordon and Breach)

### *Annual review*

*Review of Progress in Quantitative Nondestructive Evaluation* (yearly) (eds D O Thompson and D E Chimenti) (New York: Plenum)

CHAPTER 2

# Fundamental theory

## 2.1 GENERAL CONSIDERATIONS

Most of the underlying theory of electrical and magnetic testing is contained in undergraduate textbooks on electricity and magnetism (e.g. Duffin, 1980) and this chapter summarizes the general aspects of this theory. Any theory characteristic of a particular method is to be found in the relevant chapter. Système International (SI) units are used throughout this book unless otherwise indicated and a list of units appears in Appendix A. Quantities characterized by both magnitude and direction are called *vectors* and are printed here in bold type. For example, the vector $F$ which represents force can be equated to its components along the $x$, $y$ and $z$ directions of a Cartesian system, i.e.

$$F = iF_x + jF_y + kF_z \qquad (2.1)$$

where $i$, $j$ and $k$ are unit vectors directed parallel to the $Ox$, $Oy$ and $Oz$ axes. Scalar quantities, such as mass and time, are independent of direction and have values of magnitude only. Together with the magnitudes of vector quantities, they are printed in normal type.

## 2.2 ELECTRICAL CONDUCTIVITY AND RESISTIVITY

The electrical conductivity $\sigma$ of a material depends on its nature, composition and structure, e.g. grain size and hardness, and also its physical state, such as temperature. At a fixed temperature, the value of $\sigma$ for a sample of homogeneous material having uniform cross-sectional area $A$ and length $l$ is given by

$$\sigma = l/RA = 1/\rho \qquad (2.2)$$

where $R$ is the electrical resistance. The units for $\sigma$ are $S\,m^{-1}$ (siemens per metre) and $\rho$ is the electrical resistivity expressed in $\Omega\,m$ (ohm metre). Table 2.1 lists values of both $\rho$ and $\sigma$ for a number of commonly used materials.

A unit of electrical conductivity often used by engineers is the international annealed copper standard (IACS): 100% IACS is equal

**Table 2.1** Commonly used materials: approximate electrical conductivities and resistivities at room temperature[a]

| Material | Resistivity ($\Omega$ m) | Conductivity (S m$^{-1}$) |
| --- | --- | --- |
| Aluminium | $2.5 \times 10^{-8}$ | $40 \times 10^6$ |
| Cobalt | $6 \times 10^{-8}$ | $17 \times 10^6$ |
| Copper | $1.7 \times 10^{-8}$ | $60 \times 10^6$ |
| Iron | $9 \times 10^{-8}$ | $11 \times 10^6$ |
| Mercury | $96 \times 10^{-8}$ | $1 \times 10^6$ |
| Nickel | $6.8 \times 10^{-8}$ | $15 \times 10^6$ |
| Platinum | $10 \times 10^{-8}$ | $10 \times 10^6$ |
| Steels | $18-100 \times 10^{-8}$ | $1-5.6 \times 10^6$ |
| Silver | $1.5 \times 10^{-8}$ | $67 \times 10^6$ |
| Tin | $1.2 \times 10^{-8}$ | $87 \times 10^6$ |
| Zinc | $6 \times 10^{-8}$ | $17 \times 10^6$ |
| Semiconductors | $10^{-5}$ to $10^{-1}$ | 10 to $10^5$ |
| Ceramics | $10^6$ to $10^{14}$ | $10^{-14}$ to $10^{-6}$ |
| Glasses | $10^6$ to $10^{12}$ | $10^{-12}$ to $10^{-6}$ |
| Nylon | $10^{11}$ to $10^{13}$ | $10^{-13}$ to $10^{-11}$ |
| Perspex | $> 10^{13}$ | $< 10^{-13}$ |
| PVC | $10^6$ to $10^{13}$ | $10^{-13}$ to $10^{-6}$ |
| PTFE | $10^{15}$ to $10^{18}$ | $10^{-18}$ to $10^{-15}$ |
| Rubbers | $10^{13}$ to $10^{16}$ | $10^{-16}$ to $10^{-13}$ |

[a]Exact values depend on structure, composition and physical conditions. Values for individual samples of semiconducting and insulating materials may vary over several orders of magnitude depending on the method of preparation and impurity content.

to the electrical conductivity of 99.999% pure annealed copper. The conductivity of this material is equal to approximately 60 MS m$^{-1}$, so there is a conversion factor of about 1.67 from MS m$^{-1}$ to IACS. For example, pure annealed aluminium has an approximate conductivity of 40 MS m$^{-1}$ or 67% IACS.

For good electrical conductors, values of $\rho$ and $\sigma$ respectively have orders of magnitude 0.01–1 $\mu\Omega$ m and 1–100 MS m$^{-1}$ at normal temperatures. Conduction takes place in these materials as a result of the motion of free electrons when an electromotive force (EMF) is applied. The value $\rho$ increases and the value of $\sigma$ reduces with a rise in temperature. With other materials, i.e. semiconductors and dielectrics (or insulators), there are virtually no free electrons and the mechanism for conduction is different, i.e. $\rho$ decreases and $\sigma$ increases as temperature rises. At room temperature, the approximate values of $\rho$ and $\sigma$ for semiconductors are respectively 10 $\mu\Omega$ m to 0.1 $\Omega$ m and 10 S m$^{-1}$ to 0.1 MS m$^{-1}$. For dielectrics, the corresponding ranges are $10^6$ to $10^{16}$ $\Omega$ m and $10^{-16}$ to $10^{-6}$ S m$^{-1}$.

## 2.3 DIELECTRIC MATERIALS

### 2.3.1 Electric field

Although a dielectric material is virtually free from the presence of conduction electrons, it can contain an electric field arising from the presence of any free electric charges, which are either positive or negative. Their values are expressed in coulombs (C). Consider two electric charges $Q_1$ and $Q_2$ located at points $A$ and $B$, respectively, where $B$ is related to $A$ by a vector displacement $r$. **Coulomb's law** indicates that the force $F$, which is also a vector quantity, acting between them is proportional to the product of the charges and the inverse square of the displacement. Using the SI convention, this is expressed as

$$F = Q_1 Q_2 r_1 / 4\pi\varepsilon r^2 \tag{2.3}$$

where $r_1$ represents a unit vector, in the direction of $r$, and $\varepsilon$ is a quantity characteristic of the medium and called the electrical permittivity (section 2.3.2). The units of $F$, $r$ and $\varepsilon$ are the newton (N), metre (m) and farad per metre (F m$^{-1}$) (equation 2.11). When $Q_1$ and $Q_2$ are both positive or both negative, the magnitude of $F$ has a positive sign and repulsion takes place; when they have opposite signs, they are attracted to one another.

If $Q_1 = Q$ and $Q_2$ is a test charge having unit value, the force acting on the unit test charge at $B$ is defined as the electric field strength $E$ and given by

$$E = Q r_1 / 4\pi\varepsilon r^2 \tag{2.4}$$

The work done in moving unit charge in opposition to an applied field $E$ by an infinitesimal displacement $\delta r = r_1 \delta r$ is equal to $\delta V$, i.e.

$$\delta V = -E \cdot \delta r \tag{2.5}$$

so that for a finite displacement $r$ we have

$$V = -\int_0^r (Q/4\pi\varepsilon r^2)\, dr = Q/4\pi\varepsilon r \tag{2.6}$$

where $V$ is defined as the potential difference between $A$ and $B$, and is a scalar quantity. $E$ is sometimes called the potential gradient, i.e.

$$E = -\text{grad } V \tag{2.7a}$$

or

$$E = -\nabla V = -[i\partial V/\partial x + j\partial V/\partial y + k\partial V/\partial z] \tag{2.7b}$$

where $i$, $j$ and $k$ are unit vectors parallel to the $Ox$, $Oy$ and $Oz$ axes, respectively. $V$ is measured in volts (V), hence $E$ is measured in V m$^{-1}$.

16                     *Fundamental theory*

An 'uncharged' dielectric material contains no free charges. The particles (e.g. molecules) constituting dielectric materials are bound together by attractive forces, some of which are electrical by nature, and the positive and negative charges contained by these particles cancel each other. The application of an electric field is sufficient to displace these charges from one another, usually by very small amounts, so that dipoles are created. Thus, when the electric field is applied perpendicular to the parallel surfaces of a slab of dielectric material, the dipoles are aligned in that direction; positive charges appear on one surface, negative charges on the other. The number of positive charges equals the number of negative charges. The signs of the charges are interchanged when the direction of the field is reversed.

### 2.3.2 Electric flux

Electric fields can be represented as lines of force originating from and terminating at electric charges. Figures 2.1 and 2.2 illustrate lines of force connecting equal and opposite charges in conductors in the

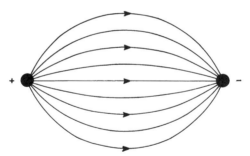

**Figure 2.1** Electrostatic lines of force between conducting spheres having opposite polarities.

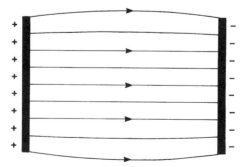

**Figure 2.2** Electrostatic lines of force between parallel conducting plates having opposite polarities.

## Dielectric materials

forms of (1) small spheres and (2) parallel plates. By convention, we can take the total number of lines (in three dimensions) as representing electric flux $\psi$, a quantity depending on the magnitudes of the charge $Q$ and the electrical permittivity $\varepsilon$.

Consider the special case of a positive charge $Q$ distributed uniformly within a spherical conductor of negligible radius about a point $O$. The lines of force are straight lines radiating from $O$ in a uniform manner in three dimensions. Defining $\psi$ as being numerically equal to the number of lines of force and specifying that one line of force radiates from unit charge, we see that $\psi = Q$ and is thus expressed in coulombs (C). Also, defining a vector quantity $D$ as follows

$$\mathbf{D} = \varepsilon \mathbf{E} \tag{2.8}$$

equation (2.4) may be rewritten as

$$\mathbf{D} = Q\mathbf{r}_1/4\pi r^2 \tag{2.4*}$$

i.e. the magnitude of $D$ is equal to $Q/4\pi r^2$. In general, for a total area $S$, we have

$$\psi = \int_S \mathbf{D} \cdot \mathrm{d}\mathbf{S} \tag{2.9}$$

where $D$ is effectively the electric flux density (cf. equation 2.15) and has the unit $\mathrm{C\,m^{-2}}$. However, for reasons which become apparent in section 2.4.2, $D$ is known as the dielectric displacement.

To obtain the values of the propagation constants of electromagnetic waves, it is necessary to obtain an expression for div $D$, i.e. the divergence of $D$. Consider an elementary rectangular imaginary prism, having dimensions $\mathrm{d}x$, $\mathrm{d}y$ and $\mathrm{d}z$, its corner nearest the origin having the Cartesian coordinates $(x, y, z)$, and containing a free charge $Q$ (Figure 2.3). The components of $D$ at $(x, y, z)$ are $D_x$, $D_y$, and $D_z$ and, at $(x + \mathrm{d}x, y + \mathrm{d}y, z + \mathrm{d}z)$, they become

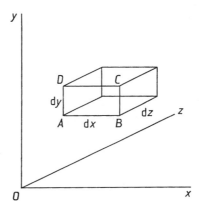

**Figure 2.3** Cartesian coordinates for an elementary rectangular prism.

$[D_x + (\partial D_x/\partial x)dx]$, $[D_y + (\partial D_y/\partial y)dy]$ and $[D_z + (\partial D_z/\partial z)dz]$. The increase in flux on passing through the prism is then equal to

$$(\partial D_x/\partial x + \partial D_y/\partial y + \partial D_z/\partial z)dx\,dy\,dz$$

In accordance with Gauss's theorem (Duffin, 1980), this is equal to the charge $Q$ contained within the prism. Putting $Q = \rho\,dx\,dy\,dz$, where $\rho$ in this context represents the charge density (in $C\,m^{-3}$), we then have

$$\partial D_x/\partial x + \partial D_y/\partial y + \partial D_z/\partial z = \rho \quad (2.10a)$$

which can be written as

$$\text{div}\,\mathbf{D} = \nabla \cdot \mathbf{D} = \rho \quad (2.10b)$$

In the absence of any free charge $\rho = 0$, so that

$$\text{div}\,\mathbf{D} = \nabla \cdot \mathbf{D} = 0 \quad (2.10c)$$

The electrical permittivity $\varepsilon$ is most conveniently defined in terms of a capacitor having parallel plates of equal area $A$ (Figure 2.2), whose capacitance $C$ is

$$C = \varepsilon A/d \quad (2.11)$$

where $d$ is the plate separation distance; it is assumed the edge effects of the plates on the field are negligible. The capacitance $C$ is defined as the ratio of the charge $Q$ on the capacitor to the potential difference $V$ across it and its unit is the farad (F) when $Q$ and $V$ are expressed in coulombs and volts, respectively. Thus the unit of $\varepsilon$ is the farad per metre ($F\,m^{-1}$).

It is usually convenient to express the electrical permittivity of a substance as a ratio, $\varepsilon_r$, known as the relative permittivity or the dielectric constant, i.e.

$$\varepsilon_r = \varepsilon/\varepsilon_0 \quad (2.12)$$

The precise value of $\varepsilon_0$, as will be seen in section 2.8.1, is directly related to the speed $c$ of electromagnetic waves in free space, very nearly equal to $3 \times 10^8\,m\,s^{-1}$, giving $\varepsilon_0 = 1/(36\pi \times 10^8)\,F\,m^{-1}$. Values of electrical permittivity for several commonly used dielectric materials are shown in Table 2.2.

Usually, for fixed conditions, the electrical permittivity remains constant for all values of $\mathbf{D}$ and $\mathbf{E}$, but there are materials called ferroelectrics in which hysteresis occurs (section 2.7).

**Table 2.2** Relative permittivities (dielectric constants) $\varepsilon_r$ and loss factors tan $\delta$ in dielectric materials at room temperature[a]

| Material | Frequency range | $\varepsilon_r$ | tan $\delta$ |
|---|---|---|---|
| Cellulose acetate | 1 MHz–1 GHz | 3.5 | 0.03–0.04 |
| Mahogany | 3 GHz | 4.3 | 0.059–0.094 |
| Marble | 1 MHz | 8.0 | 0.04 |
| Paper fibres | 50 Hz | 6.5 | 0.005 |
| Paraffin wax | 1 MHz–1 GHz | 2.2 | 0.002 |
| Polyethylene | 50 Hz–1 GHz | 2.3 | 0.002–0.003 |
| Porcelain | 50 Hz | 5.5 | 0.03 |
| Quartz | 1 kHz | 4.5 | 0.0002 |
| Silicone rubber | 50–100 MHz | 8.6–8.5 | 0.005–0.001 |
| Soda glass | 1–100 MHz | 7.5 | 0.01–0.008 |

[a]Values of $\varepsilon_r$ and tan $\delta$ depend greatly on frequency in view of dispersive effects and the best which can be done in the limited space available here is to provide some examples taken from Kaye and Laby (1973) by the kind permission of Longman Group UK Ltd.

## 2.4 ELECTROMAGNETISM

### 2.4.1 Magnetic field

When an electric field is produced, a magnetic field appears. Analogies exist between some of the properties of both these fields but, although it is possible for single electric charges to exist, it is not possible for single magnetic poles to do so. For every north pole, there is a south pole having an equal strength, and vice versa. For the sake of simplicity, the magnetic field is defined below in terms of single magnetic poles and the corresponding opposite poles are assumed to be sufficiently far removed to have negligible influence within the region of interest.

Applying Coulomb's law (equation 2.3), a force $F$ acting between two magnetic poles of strength $P_1$ and $P_2$, at points $A$ and $B$ respectively, and separated by a displacement vector $r$, is given by

$$F = P_1 P_2 r_1 / 4\pi\mu r^2 \qquad (2.13)$$

following the SI convention, where $r_1$ is a unit vector in the direction of $r$ and $\mu$ is the magnetic permeability, a characteristic of the medium. If $P_1 = P$ and $P_2$ is a unit positive (i.e. north) test pole, the force acting on the unit pole at $B$ can be defined as the magnetic field strength $H$ and expressed as

$$H = P r_1 / 4\pi\mu r^2 \qquad (2.14)$$

By analogy with electric fields, a magnetic flux $\Phi$ can be represented graphically as lines of force, although it is properly defined later (equation 2.18). For a magnetic flux threading an area **S**, by analogy with equation (2.9), we have

$$\Phi = \int_S \mathbf{B} \cdot d\mathbf{S} \qquad (2.15)$$

where **B** is the magnetic flux density, sometimes called the magnetic induction.

We also have (equation 2.8) that

$$\mathbf{B} = \mu \mathbf{H} \qquad (2.16)$$

The units of $\Phi$ and **B** are the weber (Wb) and the tesla (T) (i.e. Wb m$^{-2}$) respectively. Equations (2.19) and (2.22) indicate that **H** is measured in A m$^{-1}$ and $\mu$ in henry per metre (H m$^{-1}$). **B** and **H** are sometimes expressed in the CGS units gauss and oersted, respectively, where 1 gauss $= 10^{-4}$ T and 1 oersted $= 1000/4\pi$ A m$^{-1}$. The magnetic pole strength $P$ is seen from equations (2.14) to (2.16) to have the same unit as $\Phi$, i.e. the weber (cf. $Q$ and $\psi$ in section 2.3.2).

For convenience, the relative permeability $\mu_r = \mu/\mu_0$ is often used, where $\mu_0$ is the magnetic permeability of free space and is equal to $4\pi \times 10^{-7}$ H m$^{-1}$. The value of $\mu_r$ can be greater than unity, often by very large amounts, for ferromagnetic and ferrimagnetic materials, and it varies both with the magnetic field strength and temperature (section 2.7). The value of $\mu_r$ for other substances is very nearly equal to unity and is virtually constant for all values of **H**.

As would be expected, Gauss's theorem is applicable to magnetic as well as electric fields and, by analogy with the treatment given in section 2.3.2, we have

$$\text{div } \mathbf{B} = 0 \qquad (2.17)$$

However, the analogy of this equation with equations (2.10) is not complete because free magnetic poles are non-existent.

### 2.4.2 Interactions between magnetic and electric fields

From Faraday's law of electromagnetic induction, a change in magnetic flux $\Phi$ threading a circuit gives rise to a back EMF $V$, i.e.

$$V = -\partial \Phi / \partial t \qquad (2.18)$$

If $t$ is measured in seconds (s) and $V$ in volts (V), $\Phi$ is expressed in webers (Wb). It can be shown that the current $I$ in the circuit is proportional to the magnetic flux

$$\Phi = LI \qquad (2.19)$$

where $L$, the constant of proportionality, is defined as the coefficient of self-inductance and is measured in henries (H) when $I$ is expressed in amperes (A).

The relationship between the flux through the coil and the back EMF in the circuit is best established by using the 'curl' operator, which provides a vector in the $z$-direction as a result of action in the $x$–$y$ plane. This can be illustrated, for example, by the motion of a corkscrew. Because the significance of this operator is not always properly understood, a brief explanation is given as follows.

Consider again the elementary region in a medium in the shape of a rectangular prism (Figure 2.3) having its sides parallel to the $x$, $y$ and $z$ axes and dimensions $dx$, $dy$ and $dz$, where the coordinates of the corner $A$ are $(x, y, z)$. The potential difference between two points in a closed circuit around the path $ABCD$ in the $x$–$y$ plane and separated by an elementary displacement of $dl$ is given by the scalar product $-\mathbf{E} \cdot d\mathbf{l}$. Now the components of $\mathbf{E}$ in the $x$-direction are $E_x$ at $A$, $E_x + (\partial E_x/\partial x)dx$ at $B$, $E_x + (\partial E_x/\partial y)dy + (\partial E_x/\partial x)dx$ at $C$ and $E_x + (\partial E_x/\partial y)dy$ at $D$. It is seen that the electric field in the direction from $D$ to $C$ exceeds that in the direction from $A$ to $B$ by $-(\partial E_x/\partial y)dy$, which is the resultant field in the $x$-direction. In a similar manner it can be shown that the resultant field in the $y$-direction is $+(\partial E_y/\partial x)dx$. Hence the contribution to the line integral

$$\oint \mathbf{E} \cdot d\mathbf{l}$$

for the length of the complete circuit is equal to $(\partial E_y/\partial x - \partial E_x/\partial y)dx\, dy$ in the $x$–$y$ plane.

In the same way it can be shown that the contributions are

$(\partial E_z/\partial y - \partial E_y/\partial z)dy\, dz$   in the $y$–$z$ plane
$(\partial E_x/\partial z - \partial E_z/\partial x)dz\, dx$   in the $z$–$x$ plane

These contributions are induced by the rate of change in flux having components equal to $-(\partial B_x/\partial t)dy\, dz$, $-(\partial B_y/\partial t)dz\, dx$ and $-(\partial B_z/\partial t)dx\, dy$ along the $x$, $y$ and $z$ directions, respectively. Equating rate of change in flux to the induced EMF gives

$$\begin{aligned}-[\mathbf{i}(\partial B_x/\partial t)dy\, dz &+ \mathbf{j}(\partial B_y/\partial t)dz\, dx + \mathbf{k}(\partial B_z/\partial t)dx\, dy] \\ = \mathbf{i}(\partial E_z/\partial y &- \partial E_y/\partial z)dy\, dz + \mathbf{j}(\partial E_x/\partial z - \partial E_z/\partial x)dz\, dx \\ &+ \mathbf{k}(\partial E_y/\partial x - \partial E_x/\partial y)dx\, dy \end{aligned} \quad (2.20)$$

where $\mathbf{i}$, $\mathbf{j}$ and $\mathbf{k}$ are unit vectors in the $x$, $y$ and $z$ directions, respectively. Equating the coefficients of the unit vectors on both sides of equation (2.20) gives

$$\text{curl } \mathbf{E} = \nabla \times \mathbf{E} = -\partial \mathbf{B}/\partial t \quad (2.21)$$

where

$$\nabla \times E = i(\partial E_z/\partial y - \partial E_y/\partial z) + j(\partial E_x/\partial z - \partial E_z/\partial x) + k(\partial E_y/\partial x - \partial E_x/\partial y)$$

Using cylindrical polar coordinates it can be shown that

$$\nabla \times E = r_1[(1/r)\partial E_z/\partial \theta - \partial E_\theta/\partial z] + \theta_1[\partial E_r/\partial z - \partial E_z/\partial r] + k(1/r)[\partial(rE_\theta)/\partial r - \partial E_r/\partial \theta] \quad (2.21^*)$$

where $r_1$, $\theta_1$ and $k$ represent unit vectors.

When an electric current flows through a conductor, a magnetic field is induced in its vicinity and, for a current $I$ in a straight wire, the lines of force are circles concentric with the wire. The magnitude $H$ of the field at a point perpendicular distance $r$ from the wire is given by the Biot–Savart law

$$H = I/2\pi r \quad (2.22a)$$

It is assumed here that the wire is of infinite length. For a circular coil of radius $a$, the field is directed at right angles to the coil, and its magnitude at the centre is given by

$$H = I/2a \quad (2.22b)$$

The SI units of magnetic field are therefore amperes per metre ($A\,m^{-1}$).

A general relationship between the magnetic field strength $H$ and current $I$ can be derived by integrating over the length of a field describing a closed path consisting of vector elements $dl$ in length, i.e.

$$\oint H \cdot dl = I \quad (2.22c)$$

Equation (2.22a) is readily derived from this expression. The magnetic field in a solenoid can also be obtained from equation (2.22c), by integrating over each of its turns. If the length of the solenoid is large compared with the radius, the value of $H$ at the centre is given by

$$H = NI \quad (2.22d)$$

where $N$ is the number of turns per unit length. The field is approximately uniform over most of the length but there are divergences on approaching both ends of the solenoid. As a result, the values of $H$ at the ends are reduced to one-half the value at the centre.

Now consider conduction taking place through current elements of area $dS$. For a current density $J$, i.e. the current per unit area, equation (2.22c) becomes

$$\oint H \cdot dl = \int_S J \cdot dS \quad (2.23a)$$

However, equation (2.23a) does not take into account the effects of any break in a conducting medium as would occur, for example, in a circuit containing a capacitor. At a given time, equal and opposite charges appear on the capacitor plates, so that an electric field $E$ and a dielectric displacement $D$ (equation 2.8) are set up across the dielectric medium between these plates. The conduction current is then replaced by a displacement current $\partial D/\partial t$. Since $D$ has the units $C\,m^{-2}$, the units of $\partial D/\partial t$ and $J$ are both $A\,m^{-2}$. Hence

$$\oint H \cdot dl = \int_S J \cdot dS + \int_S (\partial D/\partial t) \cdot dS \qquad (2.23b)$$

By analogy with the derivation of equation (2.21), it can be shown that

$$\operatorname{curl} H = \nabla \times H = J + \partial D/\partial t \qquad (2.24)$$

It is often convenient (section 4.5.2) to express $B$ in terms of the magnetic vector potential $A$ defined as

$$\operatorname{curl} A = B \qquad (2.25)$$

From equation (2.21) we see that

$$E = -\partial A/\partial t \qquad (2.26)$$

## 2.5 ALTERNATING CURRENTS

### 2.5.1 Discharge and charge of a capacitor through a resistor

Consider an electric charge $Q_0$ initially given (i.e. when time $t = 0$) to a parallel-plate capacitor having a capacitance $C$ and connected across a resistance $R$ (Figure 2.4). A positive charge on one plate of the capacitor gives rise to an electric field, which induces a negative charge on the other plate, and a potential difference $V = Q/C$ appears across the plates. Thus, at any time $t$, a current $I = dQ/dt$ flows through the circuit, and the potential difference across the resistance is equal to $R\,dQ/dt$. Because there is no source of EMF, Kirchhoff's first law (section 2.6.1) shows that

$$Q/C + R\,dQ/dt = 0 \qquad (2.27a)$$

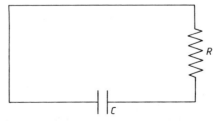

**Figure 2.4** Series $RC$ circuit.

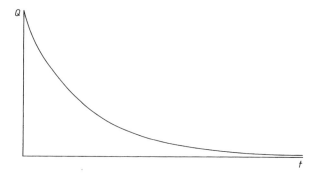

**Figure 2.5** Variation of charge $Q$ with time $t$ during discharge of a capacitance $C$ through a resistance $R$. Figure 2.4 shows the relevant circuit diagram.

The solution to this equation is

$$Q = Q_0 \exp(-t/RC) \quad (2.27b)$$

i.e. there is an exponential decay of charge with time (Figure 2.5).

The product $RC$ is known as $\tau$, the time constant or relaxation time, and is the time taken for the charge to reduce to a fraction $1/e$ of its initial value, where e is the exponential function, given as 2.718 282 to 6 decimal places.

If a direct EMF $V_0$ is inserted in series with $C$ and $R$, at any time $t$, $V_0$ is equal to the sum of the potential differences across $C$ and $R$, i.e.

$$Q/C + R \, dQ/dt = V_0 \quad (2.28a)$$

The solution to this equation is

$$Q = Q_0[1 - \exp(-t/RC)] \quad (2.28b)$$

as depicted in Figure 2.6. Equation (2.28b) shows that $Q = Q_0$ when $t$ is infinite; the capacitor is fully charged when the current is zero and $V$ equals $Q_0/C$. Here the relaxation time $\tau = RC$ is the time taken for the charge $Q$ to reach a fraction $(1 - 1/e)$ of its asymptotic value $Q_0$.

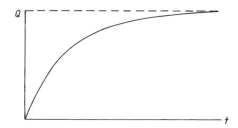

**Figure 2.6** Variation of charge $Q$ with time $t$ when a capacitance $C$ is charged through a resistance $R$.

## 2.5.2 Free oscillations of an *LRC* circuit

If an inductance $L$ is placed in the circuit in series with $C$ and $R$ without any source of EMF (Figure 2.7) and a charge $Q_0$ is given to the capacitor, it can be shown from equations (2.18) and (2.19) that a potential difference of $-L\, dI/dt$ appears across $L$ at any time $t$. Equation (2.27a) is then modified to become

$$L\, d^2Q/dt^2 + R\, dQ/dt + Q/C = 0 \qquad (2.29)$$

since $dI/dt = d^2Q/dt^2$. The solution to this equation is

$$Q = \exp[-(R/2L)^{1/2}t]\,[A\,\exp(R^2/4L^2 - 1/LC)^{1/2}t \\ + B\,\exp -(R^2/4L^2 - 1/LC)^{1/2}t] \qquad (2.30)$$

where $A$ and $B$ are constants which can be evaluated by the application of the appropriate boundary conditions.

The form of the solution of equation (2.30) depends on the relative values of $R^2/4L^2$ and $1/LC$. When $R^2/4L^2 \geqslant 1/LC$, $Q$ decreases with time in a manner similar to the discharge of a capacitor through a resistor. However, when $1/LC$ is greater than $R^2/4L^2$, damped oscillations take place and this is of interest here. The expression $(R^2/4L^2 - 1/LC)^{1/2}$ becomes an imaginary quantity and can be written as $j\omega$, where $j = (-1)^{1/2}$ and

$$\omega = (1/LC - R^2/4L^2)^{1/2} \qquad (2.31)$$

from which it can be shown that

$$Q = Q_0(\omega_0/\omega)[\exp(-\alpha' t)]\cos(\omega t - \phi) \qquad (2.32a)$$

where

$$\omega_0 = (1/LC)^{1/2} \qquad \alpha' = R/2L \qquad \phi = \tan^{-1}(\alpha/\omega_0) \qquad (2.31^*)$$

In many NDT applications we are interested in light damping, which in practice pertains to oscillations sustained for more than a few cycles, so that $R$ is sufficiently small for $\omega_0$ to be very much greater than $\alpha'$,

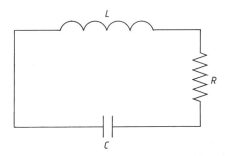

**Figure 2.7** Series *LRC* circuit.

hence $\omega$ is approximately equal to $\omega_0$. In these circumstances, equation (2.32a) effectively becomes

$$Q = Q_0[\exp(-\alpha' t)] \cos \omega_0 t \qquad (2.32b)$$

When $R$ is equal to zero the oscillations are undamped, so that

$$Q = Q_0 \cos \omega_0 t \qquad (2.32c)$$

Thus, $\omega$ and $\omega_0$ are angular natural frequencies corresponding to damped and undamped oscillations, respectively. The natural frequency $f_0'$ of the oscillations and the time period $T'$ can be related by the general expression

$$f_0' = 1/T' = \omega/2\pi \qquad (2.33a)$$

For undamped oscillations the natural frequency $f_0$ and the time period $T$ are related by

$$f_0 = 1/T = \omega_0/2\pi \qquad (2.33b)$$

The term $\alpha'$ is defined as the attenuation coefficient and $\delta = \alpha' T'$ as the logarithmic decrement (log. dec.). Here $\alpha'$ is a function of time and it should not be confused with $\alpha$, the attenuation coefficient for distance, as used with wave motion (section 2.8).

Figure 2.8 shows the variation of $Q$ with time $t$ for damped harmonic oscillations, as given by equation (2.32a), and the value of $\delta$ can be obtained from the ratio of successive amplitudes $Q_{n-1}$ and $Q_n$, i.e.

$$Q_{n-1}/Q_n = \exp \delta \qquad (2.34)$$

where for light damping

$$\delta = \alpha' T = RT/\omega L \qquad (2.35)$$

From equation (2.47) it can be seen that $\delta = \pi/\mathcal{Q}$ where $\mathcal{Q}$ is the quality factor (section 2.5.4).

**Figure 2.8** Oscillation of charge $Q$ with time $t$ during discharge of capacitance $C$ through series resistance $R$ and series inductance $L$ for values of $R^2/4L^2 < 1/LC$. Figure 2.7 shows the relevant circuit diagram.

*Alternating currents* 27

### 2.5.3 Forced oscillations of an *LRC* circuit by an alternating EMF

If an alternating EMF **V** having a frequency $\omega/2\pi$ is applied to the series *LRC* circuit shown in Figure 2.7, equation (2.29) becomes

$$L\, d^2Q/d^2t + R\, dQ/dt + Q/C = V = V_0 \sin \omega t \tag{2.36}$$

where $V_0$ is the amplitude and $\omega t$ the phase angle of **V**. Assuming the source of EMF has no internal resistance, **V** is also equal to the total potential difference across the series elements.

Equation (2.36) has two solutions. The first is the complementary function, which is obtained by putting $V = 0$ and thus provides transient values of $Q$ as given by equation (2.30). The second solution, obtained by assuming the right-hand side has a non-zero value, is called the particular integral and is characteristic of steady alternating currents. This is a complex quantity and is expressed in vector form (not to be confused with the electromagnetic vectors) by the equation

$$Q = V/(-\omega^2 L + j\omega R + 1/C) \tag{2.37}$$

The current $I = dQ/dt$ is thus given by

$$I = V/[R + j(\omega L - 1/\omega C)] = V/Z \tag{2.38}$$

where **Z** is called the complex impedance of the circuit which has a real component $R$ and an imaginary component $X$, called the reactance; the complex impedance is given by

$$Z = R + jX \tag{2.39}$$

For a series *LRC* circuit, $X = \omega L - 1/\omega C$. Defining an angle $\phi = \tan^{-1} X/R$, we have

$$Z = |Z|\exp j\phi = (R^2 + X^2)^{1/2} \exp j\phi \tag{2.40}$$

Express the sinusoidal variations of **V** with time $t$ as

$$V = V_0 \exp j\omega t \tag{2.41}$$

where $V_0$ is defined as the amplitude of **V** and $\omega t$ as the phase. Then equation (2.38) can be written

$$I = [V_0 \exp j(\omega t - \phi)]/|Z| = I_0 \exp j(\omega t - \phi) \tag{2.42}$$

where $I_0$ is the current amplitude. **I** lags in phase behind **V** by the angle $\phi$. $I_0$ is related to $V_0$ by

$$I_0 = V_0/[R^2 + (\omega L + 1/\omega C)^2]^{1/2} \tag{2.43}$$

Impedances in series and parallel can be considered in the same way as pure resistances, i.e.

$$Z = Z_1 + Z_2 + Z_3 \ldots \quad \text{(series)} \tag{2.44}$$

$$1/Z = 1/Z_1 + 1/Z_2 + 1/Z_3 \ldots \quad \text{(parallel)} \tag{2.45}$$

where the subscripts identify each impedance component.

### 2.5.4 Frequency response of an *LRC* circuit

Figure 2.9 shows the variation of the current amplitude $I_0$ with frequency in an *LRC* circuit energized by an alternating EMF maintained at a constant amplitude for all frequencies, i.e. it is the frequency response curve for the circuit. At lower frequencies it can be seen from equation (2.43) that $\omega L$ is less than $1/\omega C$ and the curve rises with an increase in frequency; at higher frequencies $\omega L$ is greater than $1/\omega C$ and the curve falls with rising frequency. The curve reaches its peak when $\omega L = 1/\omega C$, i.e. at resonance, where $\omega = \omega_0$ and $f = f_0$ for which

$$\omega_0 = 1/(LC)^{1/2} \tag{2.46a}$$

$$f_0 = 1/[2\pi(LC)^{1/2}] \tag{2.46b}$$

Equation (2.46b) is identical to equation (2.33b), indicating that the resonance frequency is equal to the natural frequency of the circuit when it is oscillating freely with minimal damping.

With an increase in the value of $R$, the curve becomes broader and shallower and its shape is characterized by the quality or $\mathcal{Q}$ factor as defined by

$$\mathcal{Q} = \omega_0 L/R = 1/\omega_0 CR \tag{2.47}$$

A comparison of equations (2.47) and (2.35) shows that $\mathcal{Q} = \pi/\delta$, where $\delta$ is the logarithmic decrement for free oscillations of the circuit. It can easily be shown that $\mathcal{Q} = \omega_0/(\omega_2 - \omega_1)$, where $\omega_1$ and $\omega_2$ are the angular frequencies on each side of the peak where $I_0$ reduces to a fraction $1/\sqrt{2}$ of its maximum, i.e. to half-power or by 3 dB (sections 2.5.5 and 2.5.6).

For a parallel resonance circuit, i.e. when $L$ and $R$ in series are in parallel with $C$, it can be shown from equation (2.46b) that the

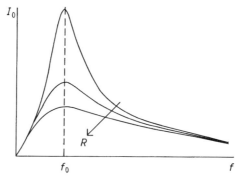

**Figure 2.9** Frequency response curves showing the effects of varying resistance $R$ in a series *LRC* circuit energized by an alternating EMF. The amplitude of the EMF is kept constant as the frequency is varied.

## Alternating currents

response curve is inverted and resonance occurs when $I_0$ has a minimum value.

### 2.5.5 AC power

The instantaneous power $P$ generated in an impedance element of a circuit is given by

$$P = V_0 I_0 \sin \omega t \, \sin(\omega t - \phi) \qquad (2.48a)$$

Integrating $P$ for a complete cycle gives the mean power $\langle P \rangle$

$$\langle P \rangle = (V_0 I_0 \cos \phi)/2 \qquad (2.48b)$$

where $\cos \phi$ is defined as the power factor. Alternating voltages and currents are usually expressed as root mean square (RMS) values, i.e.

$$V_{RMS} = V_0/\sqrt{2} \qquad I_{RMS} = I_0/\sqrt{2}$$

for which

$$\langle P \rangle = V_{RMS} I_{RMS} \cos \phi \qquad (2.48c)$$

The power consumed in a circuit containing only a pure resistor for which $\phi$ is zero is thus simply $V_{RMS} I_{RMS}$ or $(I_{RMS})^2 R$.

### 2.5.6 Decibels

The decibel (dB) scale is often useful for expressing quantities with respect to a reference level. Let $P_0$ represent a reference power and $A_0$ the corresponding amplitude or RMS value, e.g. for voltage or current. A power $P$ can then be expressed relative to $P_0$ as $10 \log_{10}(P/P_0)$ dB. Because the power is proportional to the square of both the amplitude and root mean square, the same value in dB is equal to $20 \log_{10}(A/A_0)$. In the example given in section 2.5.4 the decrease in peak current by a fraction of $1/\sqrt{2}$ of its maximum value, i.e. a reduction to half maximum power, is equal to $20 \log_{10} \sqrt{2} = 10 \log_{10} 2$, which is approximately 3 dB.

### 2.5.7 Self-inductance and mutual inductance

The concept of self-inductance was first discussed in section 2.4.2 to establish relationships between electrical and magnetic quantities, and sections 2.5.2 and 2.5.3 considered its contribution to electrical impedance. The coefficient of self-inductance $L$ of a coil having multiple turns of equal radii, i.e. a solenoid, depends on the number of turns $n$, the length $l$, and the cross-sectional area $A$; it is given by the approximate expression

$$L = \mu n^2 A/l \qquad (2.49)$$

Here $\mu$ is the magnetic permeability of the medium enclosed by the coil, i.e. its core. For an air-cored coil, $\mu$ is virtually equal to $\mu_0$ (the permeability of free space). Equation (2.49) assumes the magnetic flux is uniform throughout the length of the coil, but in practice there are divergences at its ends. However, for a long solenoid having a ratio of length to diameter greater than 5, the equation is correct to better than 1%. For a short coil of length approximately equal to its diameter, a correction factor of 0.7 should be applied.

Let us consider two coils, self-inductances placed $L_1$ and $L_2$, placed in proximity to one another but contained in different circuits (Figure 2.10). A current $I_1$ through $L_1$, the primary coil, induces a flux and thus a current $I_2$ through $L_2$, the secondary coil. The principle of conservation of energy indicates that the same current $I_2$ through $L_2$ induces a current $I_1$ through $L_1$, where $I_1$ has the same value as before.

The coefficient of mutual inductance $M$, is a function of the magnetic flux $\Phi$ linking both coils, i.e.

$$\Phi = MI \qquad (2.50)$$

where $I$ is the current induced in the secondary coil. The value of $M$ depends on the coupling factor $k$ between the coil, i.e.

$$M = k(L_1 L_2)^{1/2} \qquad (2.51)$$

and 100% coupling occurs when $k = 1$.

The voltage gain is equal to the turns ratio $n_2/n_1$ where $n_1$ and $n_2$ are the respective numbers of turns in the primary and secondary coils. Thus if $V_1$ is the potential difference across the primary and $V_2$ the potential difference across the secondary coil, the ratio of their magnitudes is given by

$$V_2/V_1 = n_2/n_1 \qquad (2.52a)$$

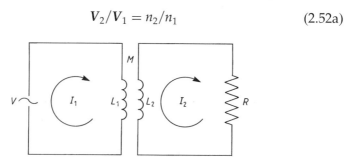

**Figure 2.10** A simple application of mutual inductance.

Potential differences are related to currents as follows

$$V_2 = j\omega M I_1 \quad \text{and} \quad V_1 = j\omega M I_2$$

Hence

$$I_2/I_1 = n_1/n_2 \quad (2.52b)$$

from which the impedance ratio is given by

$$|Z_2|/|Z_1| = (n_2/n_1)^2 \quad (2.52c)$$

## 2.6 CIRCUIT NETWORKS

### 2.6.1 Kirchhoff's laws

Circuit analysis is simplified by using Kirchhoff's laws:

1. The algebraic sum of the currents flowing towards any junction in a circuit is zero.
2. For a closed circuit, the sum of the potential differences across each element of impedance is equal to the applied EMF.

In the circuit of Figure 2.11a current $I_1$ flows in the loop *ABEF* and $I_2$ in *BCDE*, both in clockwise directions. From Kirchhoff's first law, a resultant current $(I_1 - I_2)$ flows from *B* direct to *E*. Now apply the second law.

For loop *ABEF*

$$V = I_1(Z_1 + Z_2) + (I_1 - I_2)Z_3$$

For loop *BCDE*

$$0 = I_2 Z_4 + (I_2 - I_1)Z_3$$

Substituting the values indicated in Figure 2.11b gives

$$V = I_1[R + j(\omega L - 1/\omega C)] - I_2 R$$
$$0 = I_2(R - j/\omega C) - I_1 R$$

The impedance presented in the EMF is equal to $V/I_1$, i.e. $Z'$, the components of which are

$$R' = R[1 - \omega^2 C^2 R^2/(1 + \omega^2 C^2 R^2)]$$
$$X' = \omega L - 1/\omega C - \omega C R^2/(1 + \omega^2 C^2 R^2)$$

Figure 2.10 illustrates the coupling of two circuits using a transformer. Apply Kirchhoff's second law,

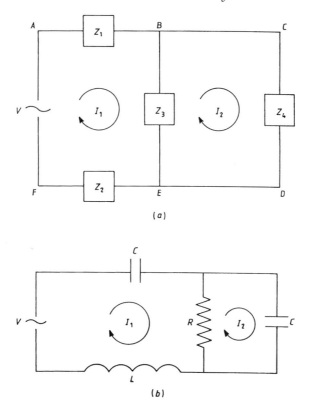

**Figure 2.11** AC networks showing series–parallel arrangements of impedance elements: (a) using complex impedances $Z_1$, $Z_2$, $Z_3$, $Z_4$; (b) using given values of resistance, inductance and capacitance. $I_1$ and $I_2$ are mesh currents.

For the primary

$$V = j\omega L_1 I_1 - j\omega M I_2$$

For the secondary

$$j\omega M I_1 = I_2(R + j\omega L_2)$$

The components of the impedance $Z'$ presented to the EMF are thus

$$R' = (\omega^2 M^2 - \omega^2 L_1 L_2)/(R^2 + \omega^2 L_2^2)$$
$$X' = \omega L_1 R/(R^2 + \omega^2 L_2^2)$$

## 2.6.2 Bridges

Impedances are often measured using a bridge (Figure 2.12a) containing four impedance arms $Z_1$, $Z_2$, $Z_3$ and $Z_4$. Balance occurs when the potential difference across the detector circuit (D) is zero, for which

$$Z_1/Z_2 = Z_3/Z_4 \qquad (2.53)$$

so that

$$(R_1 + jX_1)/(R_2 + jX_2) = (R_3 + jX_3)/(R_4 + jX_4) \qquad (2.54)$$

By separating out the real and imaginary components of this equation, two unknown impedance components can be evaluated in terms of the remaining ones, which must be known. When measuring pure resistances, a DC source of EMF is commonly used. The imaginary terms in equation (2.54) then disappear and the conditions for balance reduce to

$$R_1/R_2 = R_3/R_4 \qquad (2.55)$$

The device is known as a Wheatstone bridge.

Balance is generally obtained by making adjustments in turn to two suitable and known variable impedance components. Adjustments of

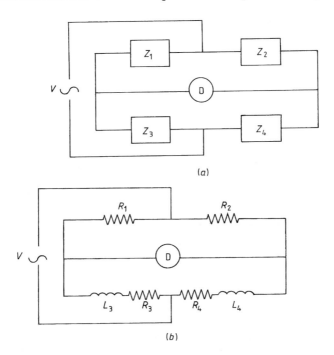

**Figure 2.12** Bridges: (a) simple bridge, AC or DC; (b) Maxwell's inductance bridge. $V$ is the applied EMF and D is the detector.

only a single impedance component need be made with the Wheatstone bridge and the basic capacitance bridge, as discussed below. A comprehensive account of the design and operation of AC bridges has been given by Hague (1934). These bridges form the basis of many types of instruments used for electrical and magnetic testing. It is important that precautions are taken to avoid unwanted signals such as those caused by interactions between the bridge components, capacitances to earth and electromagnetic radiation.

The Maxwell inductance bridge (Figure 2.12b) is a basic form of AC bridge that often appears in various forms as part of eddy current and other NDT equipment. $Z_1$ and $Z_2$ are purely resistive ratio arms, with resistances $R_1$ and $R_2$ of known and usually equal value, $Z_3$ contains the unknown inductance $L_3$, having an unknown resistance $R_3$, and $Z_4$ contains a known variable inductance and resistance, i.e. $L_4$ and $R_4$. Balance is achieved when

$$R_1/R_2 = R_3/R_4 = L_3/L_4 \tag{2.56}$$

When balance is obtained, initially under specified conditions, it is often necessary to measure changes of the inductance and resistance of a coil rather than their absolute values. For small changes of impedance, it can be shown (Blitz et al., 1981) that the values of two components, in quadrature, of the out-of-balance detector voltage are directly proportional to the inductive and resistive components of the coil.

Another basic form of an AC bridge is a capacitance bridge in which the arms 1 and 2 contain pure resistors $R_1$ and $R_2$. Arms 3 and 4 contain pure capacitors, i.e. $C_3$ being measured, and $C_4$, which is variable. Thus $\mathbf{Z}_1 = R_1$, $\mathbf{Z}_2 = R_2$, $\mathbf{Z}_3 = -j/\omega C_3$ and $\mathbf{Z}_4 = -j/\omega C_4$. Balance is achieved when

$$R_1/R_2 = C_4/C_3 \tag{2.57a}$$

If $C_3$ has a resistive component $R_3$ a variable resistance $R_4$ should be inserted in series with $C_4$ so that

$$\mathbf{Z}_3 = R_3 - j/\omega C_3 \quad \text{and} \quad \mathbf{Z}_4 = R_4 - j/\omega C_4$$

and the conditions of balance become

$$R_1/R_2 = R_3/R_4 = C_4/C_3 \tag{2.57b}$$

## 2.7 FERROMAGNETIC MATERIALS

### 2.7.1 Magnetization and hysteresis

The non-destructive testing of magnetic properties is confined mainly to ferromagnetic and ferrimagnetic materials, which can have values

## Ferromagnetic materials

of relative magnetic permeabilities $\mu_r$ (equation 2.16) ranging from unity to several thousands, depending on the nature of their properties and previous histories as well as on the magnitude $H$ of the applied magnetic field strength. For all other materials $\mu_r$ is virtually equal to unity. These materials possess either paramagnetic or diamagnetic properties, which are of fundamental scientific interest but have little or no significance to current applications of NDT. The reader who is interested in these properties and other aspects of magnetism should refer to a standard textbook on the subject (e.g. Anderson, 1968; Heck, 1974); but see section 3.8.2.

Ferromagnetism exists in a number of metals including iron, many forms of steel, nickel and cobalt. A similar phenomenon occurs in certain non-metals known as ferrites, which are said to be ferrimagnetic. These materials can be magnetized and are characterized by variations of magnetic permeability $\mu$ with magnetic field strength, generally in a non-linear manner and giving rise to hysteresis.

Consider an unmagnetized bar of ferromagnetic material placed inside and directed along the axis of a solenoid connected to a variable DC source so as to produce changes in $H$, hence in the magnitude $B$ of the magnetic flux density. Equation (2.16) can be restated as follows:

$$B = \mu_r \mu_0 H \quad (2.16a)$$

where $\mu_0$ is the magnetic permeability of free space, i.e. $4\pi \times 10^{-7}$ H m$^{-1}$.

As $H$ is increased from zero, there is a corresponding increase in $B$. Figure 2.13 shows the curve relating $B$, and thus $\mu_r$, with $H$ in a selected part of the sample surrounded by a flux-detecting coil, and $\mu_r$ is readily evaluated for all parts of the curve. At any point on the curve, the ratio $B/H$ is equal to the magnetic permeability as defined in equation (2.16). The expression $(1/\mu_0)dB/dH$, where $dB/dH$ is the gradient, is called the differential permeability $\mu_d$ for the corresponding value of $H$. For convenience, $\mu_d$ and the other expressions for permeability given in this subsection refer to the relative values, not the absolute values.

The initial curve $OAB'$ is called the magnetization, virgin or maiden curve and the value of $\mu_d$ at the origin $O$, where the curve approximates to a very short straight line, is defined as the initial permeability $\mu_i$. The ferromagnetic bar becomes fully magnetized beyond the point $A$, which has coordinates $(H_A, B_A)$, where saturation is said to occur, as indicated by the straight line $AB'$ along which $\mu_d$ is equal to unity, i.e. the relative permeability of free space. The use of saturation has proven advantageous with the eddy current testing of ferromagnetic materials because of the resulting linear relationships between the different variables, e.g. eddy current density, $B$ and $H$.

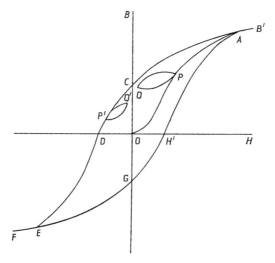

**Figure 2.13** Ferromagnetic metal: typical magnetization and hysteresis curves.

As H is decreased, the curve returns along its original path but only as far as point A from which it follows the path AC, i.e. to cut the B-axis at the point C above the origin. At point C, H is zero and B equals, $B_r$, called the retentivity or remanence, and its value is characteristic of the material. On reversing the current, and thus the direction of H, the curve continues then cuts the H-axis at point D, where B is zero and H equals $H_c$, the coercive force or coercivity, again characteristic of the material of the bar. Beyond the point E, having coordinates $(-H_A, -B_A)$, saturation occurs (EF). The value of H is then brought to zero and, on reversing its direction, it is then increased until saturation again occurs, i.e. at point A. The hysteresis loop ACDEGH'A is thus obtained. The area contained by the loop

$$\oint H \cdot dB$$

represents the energy loss per cycle for unit volume of the bar; this energy is converted into heat. A hysteresis cycle can also be obtained by passing an alternating current of suitable amplitude through the solenoid.

Magnetic hysteresis can be performed without the need to reach saturation level. For example, a common method of demagnetizing, or degaussing, a magnetized object is to subject it to a sinusoidally varying magnetic field having an amplitude which is gradually decreased to zero. The resultant curves then have progressively reducing areas. This is normally done by placing the object in the vicinity of a coil carrying a sufficiently high alternating current, usually at mains frequency, which provides the saturation value of

# Ferromagnetic materials

**Table 2.3** Magnetic properties of ferromagnetic materials[a]

| Material | $\mu_i$ | $\mu_{max}$ | $H_C$ (A m$^{-1}$) | $B_r$ (T) | $B_{max}$ (T) | $T_C$ (°C) |
|---|---|---|---|---|---|---|
| Iron (pure) | $2 \times 10^4$ | $3 \times 10^5$ | 4 | 1 | 1.5 | 770 |
| Iron (cast) | 250 | 5000 | 80 | 1.5 | 2 | 770 |
| Mild steel | 300 | 1400 | 500 | 1.2 | 1.7 | 770 |
| Nickel | 250 | 2000 | 120 | 0.3 | 0.6 | 360 |
| Piano wire |  |  |  |  |  |  |
|   Hardened | NA[b] | 1600 | 1400 | 0.8 | 1 | NA[b] |
|   Annealed | NA[b] | 5500 | 900 | 0.9 | 1.2 | NA[b] |
| Ferrites | 10–1000 | 1000–4000 | 10–2000 | NA[b] | ⩽ 0.5 | 100–600 |

[a]Orderly information about the magnetic properties of ferromagnetics is difficult to come by. Consequently, the values are approximate, abstracted from a variety of sources and must be treated only as representative. The permeabilities given are relative. $H_C$ indicates coercivity, $B_r$ retentivity and $T_C$ the Curie temperature.
[b]NA = not applicable.

$H$, and to withdraw the object slowly from the coil by a distance depending on the value of the current in it.

When magnetic hysteresis takes place in a metal subjected to an alternating field, eddy currents are induced in it. As a result, there is a further energy loss with a corresponding increase in the area of the loop (section 3.7.2). By starting at any point $P$ (or $P'$) on the curve, a minor hysteresis loop can be traced with ranges of $B$ and $H$ equal to $\delta B$ and $\delta H$, respectively. The mean gradient of this loop, i.e. the gradient of line $PQ$ (or $P'Q'$), represents the incremental permeability $\mu_{inc}$ which depends on the material properties and previous history of the material. When $P$ lies at the origin and the values of $\delta B$ and $\delta H$ are small, i.e. the minor hysteresis loop extends over the linear part of the magnetizing curve, the incremental permeability is called the recoil permeability $\mu_{rec}$. It is this value of $\mu$ which is relevant to low-amplitude eddy current testing where linearity of the variables, current, field, etc., is essential for accuracy in testing non-saturated ferromagnetic metals. The magnetic permeability decreases with rising temperature and disappears completely at the Curie temperature. Table 2.3 provides values of magnetic permeabilities at room temperature for some ferromagnetic materials.

The magnetization of a ferromagnetic material can be explained from the simplified hypothesis that the body of the material consists of a large number of **domains**, i.e. elementary magnets having microscopic dimensions, which form closed chains when the material is unmagnetized. On the application of a magnetic field, the chains progressively break down until they are oriented in a single direction when saturation occurs. The resultant changes in the magnetic flux density $B$ are not continuous but take place in discrete steps, a phenomenon known as the Barkhausen effect. These step-like changes in **B** during magnetization and hysteresis are indicated in some materials by short sharp bursts of acoustic emission.

### 2.7.2 Magnetic circuits

To minimize flux leakage from a ferromagnetic material, a closed magnetic circuit can be created. Consider the magnetization of a ferromagnetic ring of cross-sectional area $A$ (Figure 2.14) comprising four sections, each of length $l_i$, and magnetic permeability $\mu_i$, where $i$ is an integer which varies from one to four. A closed magnetic circuit containing a constant magnetic flux $\Phi$ is obtained. The corresponding magnetic fields in each section of the ring are equal to $H_i$. Taking $\Phi$ to be analogous to the electric current in a closed circuit, the magnetic equivalent of EMF is the magnetomotive force $f_m$, i.e.

$$f_m = \sum H_i l_i \qquad (2.58)$$

This equation provides an analogy with Kirchhoff's first law (section 2.6.1). From equations (2.13) and (2.14), we obtain

$$\Phi = \mu_i H_i A \qquad (2.59)$$

Completing the analogy with electrical circuits, we define a term reluctance, $R_m = f_m/\Phi$

$$R_m = \sum l_i/(\mu_i A) \qquad (2.60)$$

A relevant example of a magnetic circuit is a mild steel electromagnet ($i = 1$) with soft-iron pole pieces ($i = 2$ and 3), between which there is either an air gap, a non-ferromagnetic metal or a ferromagnetic metal ($i = 4$). To obtain a large flux, the reluctance, hence $l_i/\mu_i$, should be made as small as possible. This may be difficult when a non-ferromagnetic material lies between the pole pieces because $\mu_r$ is equal

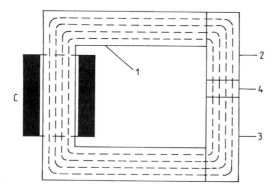

**Figure 2.14** Magnetic circuit in a magnetized ring surrounded by coil C: 1 = mild steel, 2, 3 = soft-iron pole pieces, 4 = intervening medium.

# Electromagnetic radiation

to unity, even though it may have a value of several thousands for one or more of the other components in the circuit. Hence $l_4$ must be as short as possible to provide good reluctance matching.

If a permanent magnet is used, there is no electrical excitation and the value of $f_m$ in equation (2.58) is zero.

## 2.8 ELECTROMAGNETIC RADIATION

### 2.8.1 Wave propagation in a dielectric material

For a homogeneous dielectric medium in which there are no free charges, Maxwell's equations, (2.10) (2.17) (2.21) and (2.24), can be written as follows

$$\text{div } \mathbf{D} = 0 \tag{2.61}$$

$$\text{div } \mathbf{B} = 0 \tag{2.62}$$

$$\text{curl } \mathbf{E} = -\partial \mathbf{B}/\partial t \tag{2.63}$$

$$\text{curl } \mathbf{H} = \partial \mathbf{D}/\partial t \tag{2.64}$$

where

$$\mathbf{B} = \mu \mathbf{H} \quad \text{and} \quad \mathbf{D} = \varepsilon \mathbf{E}$$

From the theory of vector analysis it can be shown for any vector $\mathbf{v}$ that

$$\text{curl curl } \mathbf{v} = \text{grad div } \mathbf{v} - \nabla^2 \mathbf{v} \tag{2.65}$$

Putting $\mathbf{v} = \mathbf{E}$ and div $\mathbf{E} = 0$ (equation 2.61), we have

$$\text{curl curl } \mathbf{E} = -\nabla^2 \mathbf{E} \tag{2.66}$$

but, from equations (2.63) and (2.64)

$$\text{curl curl } \mathbf{E} = -\mu \partial (\text{curl } \mathbf{H})/\partial t = -\mu \varepsilon \partial^2 \mathbf{E}/\partial t^2 \tag{2.67}$$

Hence

$$\partial^2 \mathbf{E}/\partial t^2 = (1/\mu\varepsilon)\nabla^2 \mathbf{E} \tag{2.68}$$

Equation (2.68) is a general equation for the propagation of unattenuated electromagnetic waves, i.e.

$$\partial^2 \mathbf{E}/\partial t^2 = c^2 \nabla^2 \mathbf{E} \tag{2.69}$$

from which the speed $c$ is given by

$$c^2 = 1/(\mu\varepsilon) \tag{2.70}$$

Putting $\mu$ and $\varepsilon$ into SI units, i.e. H m$^{-1}$ and F m$^{-1}$, the units of $c$ appear as m s$^{-1}$. Putting $\mathbf{v} = \mathbf{H}$ in equation (2.65) and continuing as before gives

$$\partial^2 \mathbf{H}/\partial t^2 = c^2 \nabla^2 \mathbf{H} \tag{2.71}$$

indicating that both electrical and magnetic fields are propagated with electromagnetic radiation.

The solutions to equations (2.69) and (2.71) for plane progressive waves travelling in the z-direction can be written in the forms

$$E_z = E_{z0} \sin(\omega t - kz)$$

and

$$H_z = H_{z0} \sin(\omega t - kz)$$

where $\omega$ is the angular frequency $2\pi f$ and $k$ the wavenumber $2\pi/\lambda$. $E_{z0}$ and $H_{z0}$ are the amplitudes of $E_z$ and $H_z$, respectively. The subscript $z$ given to $E$ and $H$ implies that the motion of the waves takes place in the z-direction. The vectors $\mathbf{E}$ and $\mathbf{H}$ lie in the plane normal to this direction. It is generally more convenient to express the sinusoidal variations as exponential functions

$$E_z = E_{z0} \exp j(\omega t - kz)$$

and

$$H_z = H_{z0} \exp j(\omega t - kz)$$

The speed $c_0$ of the waves in free space is $2.998 \times 10^8$ m s$^{-1}$. Substituting the approximation $c_0 = 3 \times 10^8$ m s$^{-1}$ and the magnetic permeability of free space, $\mu_0 = 4\pi \times 10^{-7}$ H m$^{-1}$, the electrical permittivity $\varepsilon_0$ of free space becomes equal to $1/(36\pi \times 10^9)$ F m$^{-1}$. The electrical permittivity of a dielectric material $\varepsilon$ generally has two components in quadrature, $\varepsilon'$ and $\varepsilon''$, whereby

$$\varepsilon = \varepsilon' - j\varepsilon''$$

This expression suggests (see below) attenuation of the waves and an angle is defined to provide a dielectric loss factor given by $\tan \delta = \varepsilon''/\varepsilon'$.

The solution to equation (2.68) can now be written as

$$E_z = E_{z0} \exp(-\alpha z) \exp j(\omega t - kz) \tag{2.72}$$

where $\alpha$ is defined as the attenuation coefficient and is analogous to $\alpha'$, as defined in section 2.5.2, but it refers to attenuation with distance as opposed to time. The attenuation per cycle in this context, i.e. the logarithmic decrement, can be expressed either as $\alpha\lambda$ or $\alpha'T$, where $T$ is the time period. The $\mathcal{Q}$ factor can then be expressed as

$$\mathcal{Q} = \pi/\alpha\lambda = \pi/\alpha'T$$

# Electromagnetic radiation

Substituting the derivatives of $E_z$ obtained from equation (2.72) into equation (2.68) gives

$$-\omega^2 = (\alpha + jk)^2/[\mu(\varepsilon' - j\varepsilon'')] \qquad (2.73)$$

and by equating the real and imaginary components we have

$$\alpha^2 = k^2 - \omega^2 \mu \varepsilon' \qquad (2.74)$$

as the real part and

$$k\alpha = \omega^2 \mu \varepsilon''/2 \qquad (2.75)$$

as the imaginary part. Eliminating $\alpha$ from these equations gives

$$(\omega^2 \mu \varepsilon''/2k)^2 = k^2 - \omega^2 \mu \varepsilon' \qquad (2.76)$$

i.e.

$$k^4 - \omega^2 \mu \varepsilon' k^2 - (\omega^2 \mu \varepsilon''/2)^2 = 0 \qquad (2.77)$$

so

$$k^2 = \{(\omega^2 \mu \varepsilon') \pm [(\omega^2 \mu \varepsilon')^2 + (\omega^2 \mu \varepsilon'')^2]^{1/2}\}/2 \qquad (2.78)$$

Since $k^2$ must be positive, the positive root is selected, from which it can be seen that

$$k^2 = (\omega^2 \mu \varepsilon'/2)[1 + (1 + \varepsilon''^2/\varepsilon'^2)^{1/2}] \qquad (2.79)$$

and

$$c^2 = \omega^2/k^2 = 2/\{\mu \varepsilon'[1 + (1 + \varepsilon''^2/\varepsilon'^2)^{1/2}]\} \qquad (2.80)$$

From equations (2.74) and (2.79) we see that the attenuation coefficient is given by

$$\alpha^2 = (\omega^2 \mu \varepsilon'')^2/4k^2 = \omega^2 \mu \varepsilon''^2/2\varepsilon'[1 + (1 + \varepsilon''^2/\varepsilon'^2)^{1/2}] \qquad (2.81)$$

When $\varepsilon'' = 0$, equation (2.80) shows that $c^2 = 1/\mu\varepsilon$.

### 2.8.2 Wave propagation in a conductor

A conducting material does not support an electric field but can carry currents. This means that equation (2.64) becomes

$$\text{curl } \mathbf{H} = \mathbf{J} \qquad (2.64^*)$$

where $\mathbf{J}$ is the current density. The study of electromagnetic waves through a conductor is related mainly to magnetic flux leakage and eddy current testing. The magnetic field $\mathbf{H}$ and the current density $\mathbf{J}$ are the principal variables to be considered. From equations (2.63) and (2.64*) we see that

$$\text{curl curl } \mathbf{H} = \text{curl } \mathbf{J} = \sigma \, \text{curl } \mathbf{E} = -\sigma\mu \partial \mathbf{H}/\partial t \qquad (2.82)$$

since $J = \sigma E$ (from Ohm's law). From equations (2.62) and (2.65) we have

$$\text{curl curl } \mathbf{H} = -\nabla^2 \mathbf{H} \tag{2.83}$$

Hence, equations (2.82) and (2.83) give

$$\partial \mathbf{H}/\partial t = (1/\mu\sigma)\nabla^2 \mathbf{H} \tag{2.84}$$

For plane wave propagation in the z-direction we have

$$\partial H_z/\partial t = (1/\mu\sigma)\partial^2 H_z/\partial z^2 \tag{2.85}$$

Assuming sinusoidal attenuated waves, the solution to this equation is

$$H_z = H_{z0} \exp(-\alpha z) \exp j(\omega t - kz) \tag{2.86}$$

from which

$$\partial H_z/\partial t = j\omega H_z$$

$$\partial^2 H_z/\partial z^2 = (\alpha + jk)^2 H_z$$

Substituting these quantities into equation (2.85) gives

$$j\omega\mu\sigma = (\alpha + jk)^2 = \alpha^2 + 2jk\alpha - k^2 \tag{2.87}$$

Equating real and imaginary quantities gives

$$\alpha^2 = k^2 \quad \text{and} \quad \omega\mu\sigma = 2k\alpha$$

Only the positive value of the attenuation coefficient $\alpha$ is consistent with a decrease in amplitude as z increases, hence

$$\alpha = k = (\omega\mu\sigma/2)^{1/2} \tag{2.88}$$

from which the speed $c$ of plane electromagnetic waves is given by

$$c = \omega/k = (2\omega/\mu\sigma)^{1/2} \tag{2.89}$$

These solutions can also be obtained by using $E$ as the variable instead of $H$, so they can be applied to the current density $J$ by using $J = \sigma E$.

The attenuation of the waves is often expressed in terms of the penetration or standard penetration depth $\delta$, or **skin depth**. This is not to be confused with either the logarithmic decrement (section 2.5.2) or the dielectric loss angle (section 2.8.1). It is defined as the depth $\delta$ below the surface (where $z = 0$) at which the amplitude (e.g. $H_{z0}$ or $J_{z0}$) reduces to $1/e$, i.e. approximately 8.7 dB, of its surface value. Putting $z = \delta$ in equation (2.86), we then have $\alpha\delta = 1$, so that

$$\delta = (2/\omega\mu\sigma)^{1/2} \tag{2.90}$$

Remember that equation (2.90) applies only to plane waves.

CHAPTER 3

# Magnetic methods

## 3.1 INTRODUCTION

With magnetic methods of testing, a magnetic field is applied to the object under examination and any resulting changes of magnetic flux in the region of interest are observed. Applications are confined to ferromagnetic materials and include flaw detection, measuring dimensional changes and observing variations in magnetic permeabilities brought about by changes in bulk phenomena such as hardness, grain structure and the presence of additives due to alloying and impurities. Either direct or alternating fields can be applied. However, with alternating fields, the induction of eddy currents may sometimes have to be taken into account, but this can be minimized by operating at very low frequencies, e.g. up to a few hertz. On the other hand, with some applications, the induction of eddy currents may provide useful information.

Although SI units are universally accepted, it is common practice in both the United States and the United Kingdom to retain the old CGS units for magnetic measurements, i.e. the magnetic flux density $B$ in gauss (G) and, to a lesser extent, the magnetic field $H$ in oersteds (Oe), where

$$1 \text{ Oe} = 1000/4\pi \text{ A m}^{-1} \quad (79.5775 \text{ A m}^{-1})$$

$$1 \text{ G} = 10^{-4} \text{ T } (0.1 \text{ mT}) \quad (T = \text{tesla})$$

In the CGS system, the magnetic permeability $\mu = B/H$ is dimensionless and identical to the relative permeability $\mu_r$, and the relative permeability of free space $\mu_0$ becomes equal to unity. The ratio of 1 G to 1 Oe, expressed in SI units, is equal to $4\pi \times 10^{-7}$ H m$^{-1}$. Because the CGS values of $B$ and $H$ are numerically equal to one another for free space and very nearly equal for air and other non-ferromagnetic materials, examples are often met where field strengths are incorrectly quoted in gauss, but this should not cause any confusion. Where appropriate in this chapter, values of magnetic flux density are quoted in both tesla and gauss.

Many of the recent developments in the field of magnetic testing have been carried out by Friedrich Förster and his colleagues, and the author makes no apology for devoting a large part of this chapter to describing their excellent work.

## 3.2 FLUX LEAKAGE METHODS

### 3.2.1 General considerations

Localized phenomena such as surface or subsurface cracks in ferritic steels and other ferromagnetic materials can be detected by a flux leakage method. The principal advantage is the high likelihood of detection when the method is properly applied. Another advantage is a higher sensitivity when testing for small surface cracks, even on rough surfaces, higher than any other conventional NDT method. With flux leakage methods, a magnetic field is induced inside the object being tested, and the distribution of the resultant lines of magnetic flux is determined by the values of magnetic permeability within the region of interest.

Figure 3.1 shows how discontinuities of magnetic permeability, caused by the presence of a slot simulating a defect in a magnetized ferromagnetic bar, affect the distribution of the lines of induced magnetic flux. The lines of induced magnetic flux cut through the surface, i.e. the flux 'leaks' out of the body. Associated with this phenomenon is the appearance of north and south magnetic poles on opposite sides of the slot. Flux leakage takes place not only at the surface containing the slot but also at the opposite surface, where the leaked flux densities have lower amplitudes, i.e. the leakage occurs over a wider region. Figure 3.1 also shows that some of the flux passes through the slot.

A disadvantage of flux leakage techniques is they require a sizeable component of the applied field to cut any discontinuity at right angles,

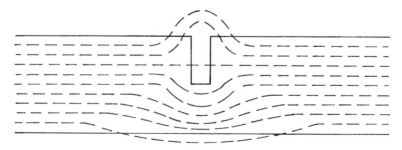

**Figure 3.1** Magnetic flux leakage at a slot cut into a magnetized ferromagnetic bar. Broken lines indicate the flux.

# Flux leakage methods

otherwise the amount of the resultant flux divergence is too small to be observed.

Flux leakage can be detected by several methods:

- Magnetic particles (section 3.3)
- Magnetic tape (section 3.4)
- Sensing coils and probes (section 3.5)

Magnetic particle inspection (MPI) is by far the most widely used flux leakage technique and the expertise gained in developing the methods of field excitation used for MPI has been applied to other flux leakage methods. Testing procedures for MPI are well covered by national and international standards (section 3.3.5).

Magnetic fields can be excited either by a permanent magnet or by electrical means with either AC or DC. AC methods are usually simpler and less expensive than DC methods but, because of the high attenuation of electromagnetic fields in metals, which increases rapidly with frequency, the penetration of alternating magnetic flux into metals is restricted (equation 2.90). DC methods provide greater sensitivities for detecting subsurface flaws but they do suffer certain disadvantages (section 3.3.3).

### 3.2.2 Demagnetizing effects

An important phenomenon associated with magnetic testing is demagnetization. When a magnetic field is applied to a ferromagnetic metal sample such as a mild steel bar, north and south poles appear near its ends and the bar becomes a magnet. The poles then produce, inside the sample, a magnetic field in a north–south direction (Figure 3.2a). Because the lines of magnetic flux density $B$ are continuous, they are in opposition to the field $H_d$ produced by these poles. The magnetic field has a maximum value in the vicinity of a pole, where there is a minimum of magnetic flux density (Figure 3.2b).

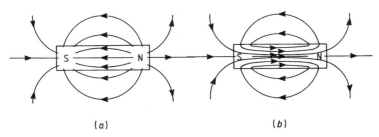

**Figure 3.2** (a) Magnetic field in a ferromagnetic bar due to the demagnetizing effects of the poles. (b) Lines of magnetic flux originally induced in the bar. (Reprinted from Stanley, 1986, by permission of the American Society for Nondestructive Testing)

The opposing field $H_d$ is called the demagnetizing field and is related to $B$ as follows:

$$B = \mu_0(M - H_d) \qquad (3.1)$$

where $M$ is the magnetic moment $m$ per unit volume of the magnet. If the axis of the bar is not coincident with the direction of the field, a torque $T$ acts on the magnet, where $T = m \times B$. The magnitude of $m$ is expressed as $m = 2Pl$, where $P$ is the pole strength and $l$ is half the distance between the poles (section 2.4.1).

### 3.2.3 Flux leakage from a slot

Consider a ferromagnetic metal ring having a rectangular section, axial circumference $l$, relative magnetic permeability $\mu_r$ and cross-sectional area $A$, with a very thin air gap of thickness $t$. When a magnetic field is applied, there is a uniform field of magnitude $H$ in the ring and a field $H_g$ in the gap, both of them coaxial with the ring (Figure 3.3). By considering the magnetic circuit (section 2.7.2), if $t \ll l$, we get

$$H_g = \mu_r H / (1 + \mu_r t / l) \qquad (3.2)$$

This result is also applicable to a gap in a magnetized rectangular bar, also provided that $t \ll l$.

Förster (1981) used this model to produce a simplified prediction of surface-crack sizes from measurements of the resultant magnetic flux leakage. Figure 3.4 shows his predictions of the lines of magnetic flux in the air spaces both above and inside the gap. Except in a very small region at the top of the gap, the magnetic field is uniform and parallel to the axis but, above the surface, the field directions trace out curves and take the form of semicircles at distances greater than $t$ from an origin located at the centre of the top of the slot. Values of the

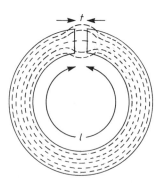

**Figure 3.3** Lines of magnetic flux in a magnetized ring, rectangular cross-section $A$ and length $l$, containing an air gap of thickness $t$.

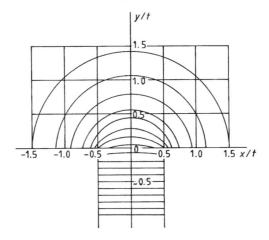

**Figure 3.4** Leakage flux lines: Förster's prediction in the neighbourhood of the slot in Figure 3.1. (Reprinted from Stanley, 1986, by permission of the American Society for Nondestructive Testing)

components $H_x$ and $H_y$ of the field at a point $(x, y)$ from the origin are given by

$$H_x = (H_g t/\pi) y/(x^2 + y^2) \tag{3.3a}$$
$$H_y = (H_g t/\pi) x/(x^2 + y^2) \tag{3.3b}$$

If the gap is then partially filled with a material identical to that of the bar, so as to provide a slot having a finite depth $d$, equations (3.3) then take the form

$$H_x = (H_g t/\pi)\{y/(x^2 + y^2) - (y + d)/[x^2 + (y + d)^2]\} \tag{3.4a}$$
$$H_y = (H_g t/\pi)\{x/(x^2 + y^2) - x/[x^2 + (y + d)^2]\} \tag{3.4b}$$

Plotting $H_x$ against $x$ using equation (3.4a) shows that $H_x$ is positive for all values of $y$, $d$ and $t$. It has been shown from flux measurements on magnetized steel samples containing slots that $H_x$ reduces, in both directions, from its peak at the centre of a slot to a negative value at or beyond the slot boundaries, depending on the value of $y$; it then increases to zero. This is caused by a demagnetizing field (section 3.2.2) which creates the appearance of free magnetic poles at the slot surfaces. The phenomenon was considered by Zatsepin and Shcherbinin who derived more realistic expressions for the components of the leakage field. A modification of their work by Förster (1982) produced the following expressions

$$H_x = (H_g/\pi)\{\tan^{-1}[A_1 d/(A_1^2 + B_1 y)] - \tan^{-1}[A_2 d/(A_2^2 + B_1 y)]\} \quad (3.5\text{a})$$

$$H_y = (H_g/2\pi)\log_e[(A_1^2 + B_1^2)/(A_2^2 + B_1^2)][(A_2^2 + y^2)/(A_1^2 + y^2)] \quad (3.5\text{b})$$

where

$$A_1 = x + t/2$$
$$A_2 = x - t/2$$
$$B_1 = y + d$$

Equations (3.5) can also be expressed in terms of flux density because the magnetic permeability of air or any non-ferromagnetic material in the slot remains effectively constant at $\mu_0$. Stanley (1986) reports that, for active field excitation, a field of 40–50 Oe (about 4000 A m$^{-1}$) can cause a leakage flux density having a value of hundreds of gauss (tens of mT) but for residual magnetism (i.e. after removal of the active field) it may only be a few gauss.

Lord and Hwang (1977) have used a finite element method (section 4.6.3) for evaluating leakage flux densities resulting from the presence of slots cut both normally and obliquely to the surface of a magnetized ferromagnetic metal; in doing so, they considered the $B/H$ curve (section 2.7.1) for the material. Figure 3.5 shows how the magnitude $B$ of the resultant flux density and the tangential and normal components, $B_x$ and $B_y$ respectively, vary with distance across the surface.

## 3.3 MAGNETIC PARTICLE INSPECTION (MPI)

### 3.3.1 General considerations

Magnetic particle inspection is the most widely used method of testing for surface and subsurface defects in ferromagnetic materials. It has evolved from the cruder method of using iron filings to demonstrate to students the existence of magnetic lines of force. With this method, a magnetic field of suitable intensity is applied to the surface of the object under test, to which minute ferromagnetic particles are applied at the same time, using either a wet or dry method (section 3.3.2). The particles line up in the directions of the magnetic flux and thus indicate flux diversions occurring at a crack or any other discontinuity of magnetic permeability (Figure 3.6a). Because the diverted lines cover a greater area than the crack, a magnified 'image' of the defect is observed (Figure 3.6b).

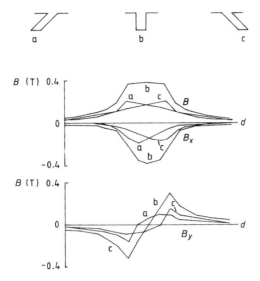

**Figure 3.5** Variations of flux density $B$ along with its tangential and normal components, respectively $B_x$ and $B_y$. The flux densities are plotted in teslas versus distance $d$ across the surface. The three types of slot are illustrated above the graphs. (Reprinted from Lord and Hwang, 1977, by permission of the British Institute of Non-Destructive Testing)

Magnetic particle inspection can be used for testing objects of virtually any size and is regarded as an essential preliminary to examining iron and steel structures and components which are likely to undergo high stresses or fatigue when in service. It is extensively used in oil and gas installations, both on land and offshore, and in the nuclear energy, aircraft and automobile industries. The method should always be used after casting, welding and heat treatment and followed, where necessary, by further tests, such as the use of eddy currents and a quantitative flux leakage method for sizing any defects which have been indicated. Magnetic particle inspection can be used to detect cracks, blowholes, laps, non-metallic inclusions and segregation, both at and immediately below the surface. The regions near the surface in which there are changes in magnetic permeability, caused by factors such as cold working and thermal disturbances, can often be indicated. The sensitivity of the method for detecting defects below the immediate vicinity of the surface (i.e. below the 'subsurface' region) is limited by the inertia of the particles.

The main advantages of magnetic particle inspection are (1) it is simple to operate, (2) it is highly sensitive, (3) it provides clearly visible indications and (4) it is relatively cheap. Although high currents are necessary to produce magnetic fields of sufficient intensity to achieve adequate sensitivity of detection, only low

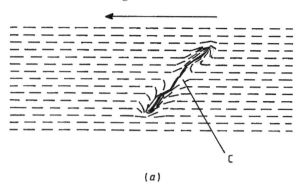

(a)

(b)

**Figure 3.6** (a) Divergence of lines of magnetic flux over the surface of a magnetized metal sample in the neighbourhood of a surface crack (C). (b) Magnetic particle indications of cracks on the surface of a hardened steel bearing; the cracks were caused by harsh grinding. (Reprinted by permission of Rolls-Royce plc)

voltages need be applied, typically 6–27 V. For example, a 6 m length of copper wire of 100 mm$^2$ cross-sectional area, used to provide the combined coil and leads, has a resistance of about 1 m$\Omega$ and requires only 1 V EMF per kiloampere. Consequently, risks from electric shock are minimal. Although MPI has the potential of provid-

## Magnetic particle inspection (MPI)

ing 100% inspection, it does depend on the operator. The method has some disadvantages:

- Only the surface and subsurface regions of objects being tested are accessible.
- Careful preparation of surfaces is required, such as the removal of grease and other substances likely to affect the mobility and visibility of the particles.
- Demagnetization must be carried out after each test. Any residual magnetism may attract swarf, which might adhere permanently, perhaps causing damage when the object is used. Residual magnetism may also present navigational problems for aircraft and ships fitted with components tested by MPI.
- Ferromagnetic particles may clog cracks, short-circuiting lines of flux and making detection harder.
- Tests are required in at least two directions because any lines of flux oriented in the direction of a defect are not diverted, thus hampering detection.
- The efficiency of the method depends on the skill, experience, concentration and integrity of the operator.
- The method could be described as messy.

In general, the maximum depth at which a subsurface crack can be detected with the use of direct fields is about 1.5 mm. With alternating fields, electromagnetic weaves are propagated into the metal but the amount of penetration is low, as characterized by the penetration depth $\delta$, typically 0.5 mm for steel tested at a frequency of 50 Hz. Equation (2.90) shows that $\delta$ decreases with increase in frequency. The use of alternating fields is usually restricted to testing for surface cracks.

The method does not normally measure defect sizes, but an experienced operator can usually estimate them roughly from the extent and pattern of the build-up of the particles at the location of a flaw. This estimation can be helped in many cases with the aid of a test block containing fine cuts having varying depths and made from the same material as the object being tested.

King (Blitz *et al.*, 1969a) has shown it is possible to identify the type of defect from the nature of the indication. For example, heat treatment cracks are usually long and narrow and are found either at an abrupt change in cross-section or at sharp contours, e.g. at the edges of holes and at the roots of screw threads. Grinding cracks are usually oriented at right angles to the direction of the grinding and severe grinding cracks are characterized by a network pattern. In castings, shrinkage cracks occur in marked changes of section. In wrought metals, laps and folds produce indications which are shorter and wider than those caused by cracks and do not follow the lines of grain flow. Inclusions of streaks of manganese sulphide are seen as lines, either continuous or intermittent, which do follow the grain flow.

Subsurface defects are usually characterized by diffuse indications that spread by amounts which increase with the depth of the defect. Other diffuse indications may indicate internal change of section, local magnetic permeability changes, caused perhaps by spot heating, resulting from a heavy current through an unprotected prod, heavy cold working, surface carburization or surface decarburization. Spurious indications may arise from local magnetization caused by rubbing against a hard pointed surface of another ferromagnetic body, but they should disappear after the object is demagnetized then retested.

The application of magnetic particle inspection to underwater testing is discussed in section 3.3.6.

### 3.3.2 Methods of detection

Either the dry or wet method may be used for detection. With dry testing, the particles consist of powdered iron filings having a high magnetic permeability and a low retentivity (section 2.7.2). They are simply dusted on to the surface of the material to be tested, usually by means of a 'pepper pot' device to give an even distribution, and blown away after the test. The dry method is suitable for testing rough castings and forgings and is particularly sensitive in locating subsurface defects. Its use is generally restricted to testing with portable equipment.

With the more commonly used wet method, the particles are usually made from magnetic iron oxide ($Fe_3O_4$) suspended in a low viscosity liquid, such as paraffin, to form an ink. The method is very suitable for surface testing, and is more sensitive and simpler to use than dry powders. It is important that the particles remain in suspension in the fluid during their application, which can take the form of immersion in a tank or bath, pouring, pumping through a nozzle, or for small-scale testing, spraying from an aerosol can. The ink should be allowed to settle down before the magnetic field is removed.

Black ink is normally used, but dyed and fluorescent inks have proven advantageous for application to dark surfaces, especially to those within apertures. Fluorescent inks are less fluid than black inks and may provide a lower sensitivity of detection for bright surfaces. Viewing of black-ink indications can be improved by initially spraying the surface with a smooth white coating. With fluorescent inks the surface is viewed under an ultraviolet or 'black' lamp. It is important the inks are kept free from contaminating agents such as oil, water and detergents. Problems pertaining to the visualization and detection of magnetic particle indications have been discussed by Chedister (1994) and the importance of the standardization of light intensity levels is the subject of a short paper by Lovejoy (1994).

A more detailed account of magnetic particle inspection is given by Lovejoy (1993).

### 3.3.3 Magnetic field excitation

The magnetic field can be generated by one of the following methods:

- Placing a magnetized yoke in contact with the object
- Passing a current through the object
- Magnetic induction, using either a coil or threading bar

The direction of the applied magnetic field should make a sufficiently large angle with any defect or other discontinuity of magnetic permeability so as to provide the maximum sensitivity of detection. This is best achieved by conducting two successive tests; in the second test the magnetic field should be at an angle of 90° to the field in the first test. This ensures a sizeable field component at right angles to any discontinuity (Figure 3.6a). Different techniques may often have to be used for each of these tests. The structure of the object being tested may sometimes make it difficult to achieve an adequate field component in all of its parts and it may be necessary to apply three fields in selected directions.

Using modern equipment, multiple tests can be made simultaneously using either two separate AC excitations at the same frequency or with a mixture of AC and DC excitations. Deutsch and Vogt (1982) have shown that, with two simultaneous AC excitations differing in phase by 90°, the resultant field vector is an ellipse which rotates at the same frequency as the current, so that once per cycle its major axis lies perpendicular to the line of any crack which is present. The required accumulation of magnetic particles over the crack is thereby attained. Simultaneous excitations from an AC coil and a DC yoke produce a linear field which performs a simple rotation at the AC frequency. Three-phase AC can be used to provide magnetizations in three different directions.

The maximum sensitivity of detection is obtained for values of $H$ in the region of inflexion of the $B/H$ curve (section 2.7.1), i.e. where the magnetic permeability is maximum. A high field can produce indications of surface roughness in the form of noise, and thus a reduction in sensitivities of detection. Too high a field may cause saturation and the magnetic permeability is consequently reduced to an unacceptably low value. With some AC applications this may be an advantage for detecting subsurface cracks because, although saturation may occur at the surface, the decrease in magnetic field with penetration might be sufficient to produce maximum permeability in the subsurface region.

The choice of magnetic field strength depends on the value of the magnetic reluctance, which is governed by the size and magnetic permeability of the object being tested and of any metal pieces used either to complete a magnetic circuit or to maintain parallel flux lines in the region of interest (section 2.7.2). More details of the choice of frequency are provided by the appropriate standards (Appendix B).

The active method is normally used, where the particles are applied during the excitation of the field, but the residual field method may sometimes be more suitable. In the residual field method, the particles are applied after the excitation is removed. The active method sometimes makes it difficult to distinguish between surface and subsurface defects on the basis of the particle indications, but if the test is repeated with only a residual field present, there is little or no indication of the presence of subsurface defects.

Although a permanent magnet may produce a field of adequate strength for some magnetic particle tests, it is more common to apply an electric current of the order of thousands of amperes to magnetize the test sample. To avoid the consequences of overheating, the energizing times should be as short as possible. A cheap and highly effective power source is an AC mains transformer which can supply currents exceeding 25 kA. High currents maintained for only very short periods can be generated by discharging capacitors (section 2.5.1) and Moake and Stanley (1985) have reported the application of this method with the use of pulsed magnetic fields. The method has also been used for magnetographic underwater testing (section 3.3.6).

The method of DC excitation is preferable for detecting subsurface defects. If mains electric power supplies are unavailable, heavy-duty rechargeable batteries can provide alternative sources for inducing direct fields. However, they are expensive to buy and to maintain, have relatively short lives and are unable to sustain high currents over the long periods required for extensive testing. When an electrical mains supply is available, rectified high-ratio current transformers are normally used for DC excitation. These DC methods suffer the disadvantages of overheating when metal clamps are placed in contact with the test sample. Stray fields may appear in the vicinity of each clamp, masking any defect indications.

*Yoke methods*

A yoke used for field excitation can take the form of either a permanent magnet or an electromagnet, operated by either AC or DC. It is shaped in such a way that, when placed in contact with the object under test, a complete magnetic circuit is formed (section 2.7.2). The induced field is longitudinal in direction and thus suitable for detecting transverse defects. To keep the reluctance to a minimum, precautions should be taken to ensure there are good contacts between the surface of the test piece and the pole pieces of the yoke; if necessary, this can be done by grinding the surfaces flat. Good contacts with curved surfaces are obtained by using suitably shaped pole pieces.

The simplest type of yoke is the permanent horseshoe magnet. This is easy to manipulate and, when made from a high-retentivity ferromagnetic alloy, it can produce very high magnetic fields. A typical example is made from an aluminium–nickel–cobalt alloy, weighing 2.34 kg, having dimensions 92 mm × 125 mm × 47 mm and producing a field of 120 kA m$^{-1}$ at the centre of an air gap 68 mm wide between the poles.

Modern permanent magnets used for magnetic particle inspection can be very flexible in their design and some are fitted with hinged arms and rotating pole pieces (Figure 3.7). If the pole pieces are rigid and the object being tested is shorter than the distance between the poles, the magnetic circuit can be completed by placing a suitable piece of mild steel in series with it (Figure 3.8).

Permanent magnets have the advantages of cheapness, convenience, portability and easy manipulation, but they can be used for testing only small areas at a time. Variation of the magnetic field strength to obtain a suitably high value of permeability can be made only in a downward direction, i.e. by increasing the reluctance of the circuit by interposing thin pieces of a non-ferromagnetic material, such as aluminium, between the pole pieces and the test object. Permanent magnets, nevertheless, are highly suitable for rapid spot

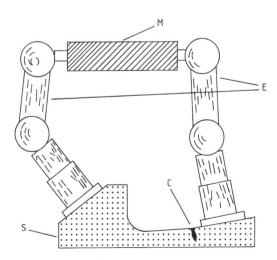

**Figure 3.7** Permanent magnet (M) with mild steel hinged extensions (E) for energizing a field in a test sample (S) containing a crack (C). (Reprinted by permission of Rolls-Royce plc)

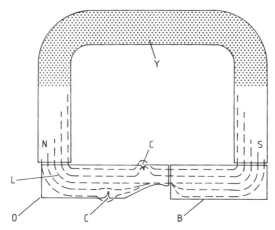

**Figure 3.8** Horseshoe permanent magnet yoke (Y) for testing an object (O) containing a crack (C) using a mild steel block (B) to complete the magnetic circuit of flux lines (L). (Reprinted by permission of Rolls-Royce plc)

checks on castings and welds and can be used to great advantage underwater. However, they have a tendency to demagnetize themselves and, for this reason, the magnetic circuit should be kept closed by means of a keeper when not in use.

Better results can often be obtained by winding a coil round a mild steel yoke to form an electromagnet (Figure 3.9). The coil should be made from heavy-duty cable because of the high current required to produce a direct or alternating magnetic field of the required strength, i.e. $10\,\text{kA}\,\text{m}^{-1}$ or more. This level can be attained by winding a few

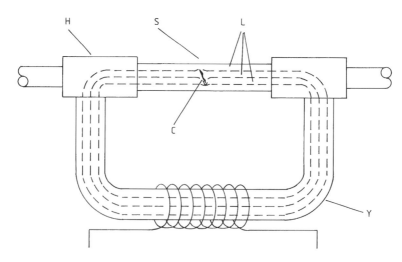

**Figure 3.9** Electromagnet yoke (Y) with adjustable contact head (H): C = crack, L = flux lines, S = test sample. (Reprinted by permission of Rolls-Royce plc)

turns of wandercable around the yoke to form a coil and passing a current of about 1 kA. The magnetic field intensity can be varied simply by adjusting the value of this current. Some commercially available yokes, which may be more convenient to use, are excited by a larger number of windings of light-duty wire, carrying a current of only 2 A, but this introduces a high inductive impedance, which leads to a reduction in field strength.

The sensitivity of detection for any portion of a component being tested varies with the area of cross-section, hence the magnetic reluctance, as stated earlier. For example, a given magnetic field intensity which can produce a maximum permeability in one region might in another region produce saturation, where $\mu_r$ approaches unity. Hence, a sample having a varying cross-section should be repeatedly tested in different positions with appropriate values of magnetic field intensity so as to ensure the maximum attainable values, hence the optimum sensitivities of detection in all of its parts. Because the reluctance increases with length, the yoke method is restricted to testing sections less than about 300 mm long and is thus suitable only for small objects and small areas of larger objects.

### Current flow methods

Passing a current through a material provides a straightforward means of inducing a magnetic field in it. The current flows in an axial direction and the magnetic flux lines are closed curves in planes at right angles to the axis, in accordance with the Biot–Savart law (equations 2.22), thus allowing the detection of defects lying in directions having sizeable components parallel to the direction of the current, i.e. perpendicular to the field. Excitation is usually made by direct contact, i.e. by the contact current flow method, and in its simplest form the test object is located between two pole pieces made from massive copper blocks faced with copper gauze (Figure 3.10). Gauze is used because it helps to reduce the risk of damage from overheating and sparking because of the high currents carried. The current should be monitored continuously with an ammeter to avoid the possibility of contact failure, which reduces the magnetic field strength.

When part of a large surfaced is tested, the current can be conducted to and from it through a pair of point prods (Figure 3.11) or G-clamps connected to the current source by wandercables. Note that, for a given applied voltage, the longer the cable, the greater its resistance, hence the heavier the loss in current. Again precautions should be taken to avoid overheating and sparking because, for very high currents, too small an area of contact can lead to serious damage to the test sample.

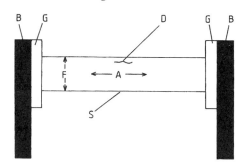

**Figure 3.10** Current flow method for testing a steel bar (S): A = direction of current, B = copper block, D = defect, F = direction of flux, G = copper gauze. (Reprinted by permission of Rolls-Royce plc)

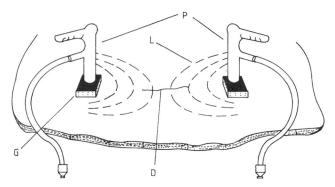

**Figure 3.11** Prod testing with wandercables: D = defect, G = copper gauze, L = lines of flux, P = metal prods. (Reprinted by permission of Rolls-Royce plc)

An important feature of using point prods is that the current paths are elliptical (Figure 3.12). Complete coverage of the area of interest is ensured by having these prods scan the surface (Figure 3.13) at intervals of half the prod separation (i.e. for an elliptical axis ratio of 2:1). The scanning intervals in the direction perpendicular to the straight line connecting the prod contact points depend on the value of the current. An increased current allows fewer intervals and thus gives the advantage of a greater elliptical curvature consistent with an adequate magnetic field. The upper value of current may be limited by the need to avoid magnetic saturation in the regions where the field is a maximum, as well as by the risks of overheating caused by the prod contacts. This method is especially suitable for scanning very large surfaces such as the hull of a ship. A fuller account of this method is given by King (Blitz et al. 1969a).

The use of the induced current flow method, ideal for detecting circumferential defects in large rings, has the advantage of avoiding

## Magnetic particle inspection (MPI)

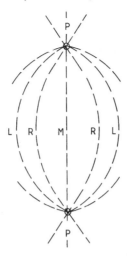

**Figure 3.12** Current flow lines from prods (P): maximum (M), reduced (R) and low (L) currents. (Reprinted by permission of Rolls-Royce plc)

difficulties with contacts. The current is induced in the sample by a transformer (Figure 3.14) in which part of the core can be detached to allow positioning of the sample to form the secondary circuit.

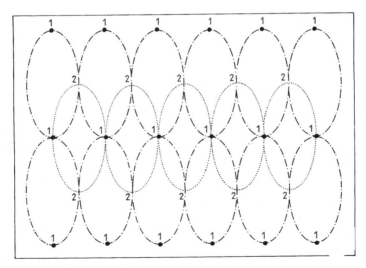

**Figure 3.13** Elliptical fields for prod testing: major axes of ellipses 2 are displaced laterally from those of ellipses 1 by half the prod separation. (Reprinted by permission of Rolls-Royce plc)

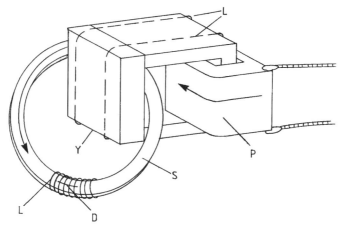

**Figure 3.14** Induced current flow method for testing a steel ring: D = defect, L = flux lines, P = transformer primary, S = steel ring, i.e. transformer secondary, Y = yoke. (Reprinted by permission of Rolls-Royce plc)

*Field induction methods*

A magnetic field, either direct or alternating, can be induced in the object under test by passing a current through a conductor in its vicinity. The value of the induced field is related to the current in accordance with the Biot–Savart law (equations 2.22). The usual technique is to wind a coil around the test object so the field lies in the direction of the axis of the coil. The coil can either be rigid and made from heavy-duty copper strip or flexible and made of cable windings. Alternatively, the conductor may consist of one or two straight wires or rods as, for example, with the threading bar. Induction methods often complement current flow methods; for example, when testing a bar, longitudinal defects are indicated by current flow and transverse defects by induction methods. There is a danger of demagnetization (section 3.2.2) when testing a short object but this can often be prevented by extending its length effectively with a steel bar in close contact with it, thus ensuring a uniform field in the region of interest (Figure 3.8).

The threading bar is a copper rod which can be inserted through a tube and can be connected at its ends by a heavy cable to the current source so as to produce coaxial circular fields at both the outer and inner surfaces (Figure 3.15). A higher degree of sensitivity is achieved by locating the bar close to the inner surface, thus increasing the magnetic field strength, and also by rotating the tube in an eccentric manner and scanning it; care should be taken to avoid contact between the rod and the sample. Greater coverage and uniformity of the field can be achieved by using two or more bars connected in parallel.

## Magnetic particle inspection (MPI)

When testing large-diameter tubes, hollow threading bars are sometimes used. Threading coils are effective for inspecting cracks in bolt holes located close to the edge of a steel sheet. A single turn suffices for a 25 mm diameter hole. Figure 3.16 illustrates a threading coil used for testing a steel ring.

Although the choice of methods of excitation depends on the shape and size of the object under test, it is essential to find all defects, irrespective of direction. Table 3.1 indicates the optimum defect orientations for the various methods of excitation.

### 3.3.4 Demagnetization

For the reasons given in section 3.3.1 it is usually necessary to demagnetize an object after any kind of flux leakage test. The most

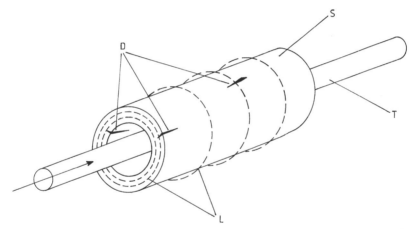

**Figure 3.15** Using a threading bar (T) to test a small-diameter tube: D = defect, L = lines of flux, S = tube. (Reprinted by permission of Rolls-Royce plc).

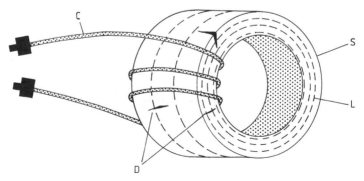

**Figure 3.16** Threading coil (C) for testing a steel ring (S): D = defect, L = flux lines. (Reprinted by permission of Rolls-Royce plc)

**Table 3.1** Magnetic particle testing: optimum directions of detectable defects for different methods of excitation

| Method | Defect direction |
| --- | --- |
| Yoke | Transverse to direction of field |
| Contact current flow | Longitudinal to direction of current |
| Induced current flow | Circumferential in rings |
| Field induction | Transverse to axis of coil |
| Threading bar | Longitudinal to direction of bar |

effective way of doing this is first to magnetize it with a field exceeding the level of the coercivity $H_c$ (section 2.7.1) then to subject it to hysteresis cycles having amplitudes that progressively decrease to zero. This is most easily done by exciting an encircling coil with a sufficiently high alternating current then withdrawing the object slowly until it is beyond the influence of the magnetic field. This technique is highly effective for testing components on a production line, but when its use is not feasible, the current can be reduced to zero while the object remains in the coil. With large objects, the use of AC may not be able to supply a sufficiently high degree of flux penetration and it is better to obtain hysteresis by varying a direct field slowly in a cyclical manner with a decreasing amplitude, i.e. effectively applying an alternating field at a frequency of a fraction of a hertz.

### 3.3.5 Testing procedures

The magnetic particle method of inspection is long established and widely used. It is capable of locating and assessing defects which, if not detected, could lead to a catastrophe with possible loss of human life. The degree of confidence in the method is so high that a component which passes a magnetic particle test is often allowed to go into service without any further testing. Well-defined specifications have been laid down by both international and national standards organizations, e.g. the International Standards Organization (ISO), the British Standards Institution (BSI) and the American Society for Testing and Materials (ASTM), for the selection of methods, the maintenance of equipment, the choice of consumables, testing procedures and the training, certification and physical well-being of personnel (Appendix B). Strict controls are required for degrees of concentration of magnetic inks, cleaning and demagnetizing surfaces, values of exciting currents and the careful monitoring of their levels along with the intensity of ultraviolet lamps, when used, because of any accumulation of dirt and deterioration with age (Nash, 1977). Frequent periodic checks on

## Magnetic particle inspection (MPI)

the efficiency and proper functioning of the equipment are necessary and specifications are laid down for the correct preparation of test blocks made from the materials to be examined.

It is mandatory to make a careful record of all tests, including details of the equipment, test blocks, directions of fields, values of magnetizing currents, the nature of the inks used, methods of viewing, etc. Most important, however, is to keep a permanent record of all defect indications. King has listed the following different ways of doing this.

- Allow surplus ink to drain off and coat the indications with clear lacquer.
- Cover the indications with clear sticky tape (e.g. Sellotape).
- Pick up the indications with clear tape and transfer them to white paper.
- Cover the indications with cold-curing plastic (e.g. PVC paste) and cut off the replicas when set.
- Photograph the indications, preferably using the same illumination conditions as for viewing.

Because the magnetic particle method is highly dependent on the skill of the operator, it is essential that personnel should be given the best possible working conditions. When using fluorescent inks, the operators should be allowed sufficient time to acclimatize themselves to the viewing conditions and should always be given regular periods of rest to counteract fatigue. Eyesight should be regularly monitored.

When carried out manually, magnetic particle inspection can prove to be both time-consuming and expensive because of the need to employ highly skilled operators. Any attempt to increase the scanning speed reduces the reliability by up to 30%. However, the introduction of a suitable and properly designed automatic system should decrease the speed of scanning and hopefully achieve 100% reliability. Pautz and Abend (1996) have considered the setting up of such a system for which the illuminated test surface is scanned by a computerized digital camera while maintaining a constant magnetic field strength and a stabilized particle concentration in the penetrating ink. The system is initially calibrated to relate its output to the characteristics of any defects encountered. Information on the position, size and nature of such defects is printed out.

During a test it is essential that the characteristics of the system remain constant. Thus if any unwanted variations occur in the ink concentration, the strength of the applied magnetic field or the intensity of illumination, the test should stop automatically; it should not be resumed until the system has returned to normal.

### 3.3.6 Underwater applications of magnetic particle inspection

Magnetic particle inspection (MPI) is extensively used by the oil and natural gas industries for testing underwater marine structures such as platform supports, pipelines and risers, especially for weld defects. The testing personnel are qualified divers who may have to operate at depths of the order of 150 m. Most of the defects likely to require MPI are surface-breaking fatigue cracks, hence AC methods are generally used. The structures are usually covered with marine organisms and, before testing, they must be cleaned down to bare metal, usually by high-pressure fluid jets.

The flow of the particles is inhibited by the pressure exerted by the sea and it may be necessary to increase the magnetic field strength; Hatlo (1979) suggests a value of 25% above the values for similar tests carried out above the water. With the aid of compressed air, the particles are generally applied through a hose fed from a reservoir. The hose should be as short as possible to allow the correct particle concentration to be maintained. Although underwater visibility is usually good, it decreases in the vicinity of the sea bed as a result of swirling mud, so fluorescent inks are advisable.

A widely used method of magnetization is the current flow method using a pair of prods (Figure 3.11); a current of about 5 A per millimetre spacing provides an adequate field intensity. The prods are capable of operation over surfaces having a high degree of curvature, characteristic of tubular joints. They are rigidly mounted on a device which also contains the ink applicator, so the diver can operate them using only one hand.

The use of AC yokes is possible for plane surfaces provided that good contact with the pole pieces is assured. To avoid power losses, the current leads should be as short as possible and, to ensure this, the transformer should be close to the operator. It is essential that both transformer and energizing coils are properly protected against the effects of sea water.

Another method especially suitable for testing welds works on a similar principle to the two parallel threading bars of section 3.3.3. A magnetic field is induced by passing currents, both in the same direction, through two parallel cables located close to the weld, so the whole area of interest can be examined in a single operation.

Permanent magnets have been widely used in the past because of their comparative simplicity. Unfortunately, they have proved in many cases to be unreliable for finding defects.

An important feature of underwater testing is to obtain reliable records of defect indications, which is best achieved with ultraviolet flash photography (Mills *et al.* 1983). This is not always feasible and one alternative is to use magnetic putty instead of ink. Containing high-retentivity iron particles, this putty is applied to the surface and then suitably shaped. After the field has been applied, the putty is removed and taken to a convenient location above the water, where magnetic ink is applied. A better but more expensive alternative is the use of magnetography (section 3.4).

As pointed out by Groves and Connell (1985), because of the high expenses incurred with marine technology, it is important that the divers employed as operators are both proficient and reliable. Further information on the use of magnetic particle inspection underwater has been given by Hatlo (1979).

## 3.4 MAGNETIC TAPE INSPECTION (MAGNETOGRAPHY)

A possible alternative to magnetic particle inspection is the use of magnetic tape, i.e. magnetography. It has the advantages of providing a permanent record, being independent of the qualities of the operator, and thus highly suitable for underwater detection. The method is clean and is readily adaptable to automation. It has the distinct advantage of being quantitative, i.e. it can measure crack depths; lower magnetic fields are sufficient for the detection of surface cracks.

In principle the method consists of magnetizing the test sample using one of the methods described in section 3.3.3 and covering the part of the surface to be examined with a length of magnetic tape, which can later be scanned by a probe (section 3.5).

Förster (1983) describes a magnetic tape method suitable for testing steel billets on a production line. A billet (B) (Figure 3.17) moves horizontally at a steady speed of about $1 \text{ m s}^{-1}$ through the region of an AC induced magnetic field; the top surface of the billet is in contact with the lower part of an endless loop of magnetic tape (T) which picks up leakage fields corresponding to any defects present. A suitably located eraser wipes off the signals from the tape after they have been received by the probes. The tape is scanned at a rate of 250 times per second by a rotating disc (P) containing several probes which are equally spaced on a circumference coaxial with the disc. Each probe is linked to a colour spray forming part of an array (G) and the signals are monitored by an indicator (I). Marks of different colours indicating levels of flux leakage detection thus appear on the surface of the billet at positions of defects. For example, a red mark might indicate a defect large enough to require rejection; a white mark might indicate a defect small enough to be harmless.

For offshore testing, the primary advantage of the magnetic tape method over the use of magnetic particles is that no measurements need be conducted underwater and the diver is not required to have any knowledge of non-destructive testing techniques. Tapes of 0.5 or

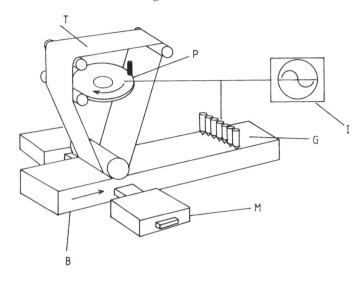

**Figure 3.17** Magnetography testing of a steel billet (B): M = motor, P = contactless pick-up containing probes, T = magnetic tape. (Reprinted from Förster, 1983, by permission of the American Society for Nondestructive Testing)

1 m in length and from 50 to 100 mm in width are placed on the test object, e.g. along the length of a pipe, and kept in a position by permanent magnets. Using a technique developed by Förster (Stumm, undated), a short magnetic pulse of a few tens of milliseconds duration is generated from a capacitor charged by a battery to a flexible yoke having suitably shaped pole pieces and located over the area of interest. Because of lower power requirements, as compared with those for magnetic particle inspection, the battery has sufficient energy to supply several hundred pulses. The tape can be scanned above water in a suitable environment by a skilled operator using a sensitive detector such as a Hall element (section 3.5.2), either singly or in an array and usually at high speed. The record on the tape is permanent, therefore scans can be repeated as often as required and with different signal treatments. The tape is subjected to a C scan, i.e. a scan covering a complete $x-y$ plane so as to indicate contours of selected output levels. This produces patterns similar to those obtained with magnetic particles.

An important characteristic of the method is that, because tapes are flexible, sharply curved surfaces, as found at welds in the T-junctions of pipes, can be examined by scanning a flat length of tape. Although the cost of magnetographic equipment is very high compared with conventional magnetic particle apparatus, there are great savings in that a highly paid diver, unhampered by heavy equipment and the

*Quantitative flux leakage detectors* 67

need to make observations, can increase his or her speed of operation by a factor of up to 5, and a single set of detecting equipment can simultaneously service a large number of sites.

## 3.5 QUANTITATIVE FLUX LEAKAGE DETECTORS

### 3.5.1 General considerations

Magnetic particle inspection, as used at present, does not provide a properly quantified means of assessing flux leakage. However, there are several surface-scanning probes capable of measuring values of magnetic flux density, which can be related to defect sizes if a preliminary calibration is made. The probes may be classified under the following headings:

- Hall probes
- magnetoresistive sensors
- coils
- SQUIDs

Compared with magnetic particles, these devices are all more sensitive to variations in flux density. They can be used to scan the surface of a test object at any desired degree of lift-off, i.e. height above the surface, and can measure both vertical and horizontal components of flux density with suitable axial orientations. They can be used singly, with either manual or automatic scanning, or mounted in arrays containing typically 12 or more elements arranged in a line perpendicular to the direction of the scan. As mentioned in section 3.2.1, the methods of field excitation are generally the same as those used for magnetic particle inspection (section 3.3.3), but modifications consistent with large-scale automatic testing are often made (section 3.6).

### 3.5.2 Hall probes

A Hall probe is essentially a rectangular slab of conducting material, i.e. the Hall element, placed in a magnetic field to measure flux density. Figure 3.18 illustrates such a slab *PQRS* having a width $w$ and a thickness $t$ with a current $I$ flowing along its length. If a magnetic flux density of magnitude $B$ is directed at right angles to the surface, a potential difference $V$ appears across the width, given by

$$V = R_H J B w \qquad (3.6a)$$

Here $J$ is the magnitude of the current density and is equal to $I/wt$, thus

$$V = R_H I B / t \qquad (3.6b)$$

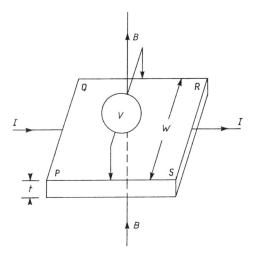

**Figure 3.18** Hall element: $B$, $V$ and $I$ are respectively the magnetic field, the current and the potential difference; $w$ and $t$ are the width and the thickness of the element.

where $R_H$ is defined as the Hall coefficient, which is constant for a given material at a fixed temperature and is expressed in $m^3 C^{-1}$. Hence, for a specified current and temperature, $V$ is proportional to $B$. The effect occurs for both direct and alternating fields, and with alternating fields the value of $V$ varies at the same frequency as $B$, provided the current $I$ is DC.

The Hall effect occurs for all conducting materials and it can be shown for metals (Wright, 1966) that

$$R_H = 1/ne \qquad (3.7)$$

where $n$ is the free carrier density, i.e. the number of electrons per unit volume, and $e$ the charge on the electron, equal to $-1.6 \times 10^{-19}$ C. For metals, $n$ is of the order of $10^{26}$ m$^{-3}$, for which $R_H$ has a magnitude of $10^{-7}$ m$^3$ C$^{-1}$, far too small to be of any practical use. Semiconductors, on the other hand, are more suitable for the design of Hall probes and many of them can be **doped** with impurities to provide conducting carriers in the forms of either **electrons** or **holes** (i.e. positive carriers) so as to give a suitable value of $n$. The expression for $R_H$ for semiconductors is given approximately by equation (3.7). The choice of semiconducting material is governed by the need for stability of the Hall effect at room temperature and a low temperature coefficient of $R_H$. Indium arsenide meets these requirements very well.

The maximum sensitivity $V/B$ of a Hall probe is achieved when $R_H$ is large and the thickness $t$ of the element small (equation 3.6b). The spatial variations of $B$ in the vicinity of a defect are high and the

## Quantitative flux leakage detectors

surface area of the probe should therefore be as small as possible. Typical dimensions of a Hall element for flux leakage measurements are 1 mm long, 0.5 mm wide and 0.05 mm thick. The use of smaller dimensions presents difficulties with the proper locations of the electrical connections. The potential difference connections should be centrally positioned and, to allow a uniform flow in a perpendicular direction in the absence of a magnetic flux, the current electrodes should completely cover the sides $PQ$ and $RS$ of the element (Figure 3.18). The effect of the flux is to distort the current paths so as to create the appearance of the potential difference $V$. Overheating the element is avoided by passing only a small current, e.g. 1 mA. For indium arsenide, equation (3.7) shows that a free electron carrier density of $10^{18}$ m$^{-3}$ produces a Hall coefficient of about $-6$ m$^3$ C$^{-1}$. Thus for a current of 1 mA, the sensitivity $V/B$ is about 100 V T$^{-1}$, i.e. a measured potential difference of 1 µV corresponds to a flux density of 10 nT (i.e. $10^{-4}$ gauss).

Manufactured probe assemblies range from single portable devices with visual indications to automated units with large-scale multiple-channel systems containing probe arrays. Automated units decrease the duration of scanning. Flux densities down to 100 µT can be measured without difficulty. More sensitive detecting equipment can provide a measuring range extending from 0 to 100 nT, with a resolution better than 1 nT, as used with some of the applications described in sections 3.4 and 3.6.

Hall probes made from suitable materials have the advantages of high degrees of sensitivity, small sensing areas and the ability to operate at high temperatures. They are not as robust as coil sensors and thus require encapsulation, which can give rise to a relatively large amount of lift-off from the inspected surface. With differential measurements, the sensitivities of each of the probes should be identical. In practice this can be achieved by incorporating microcircuits into the bodies of the probes and suitably adjusting their characteristics.

### 3.5.3 Magnetoresistive sensors

With the phenomenon of magnetoresistance, which is related to the Hall effect, a fractional increase in the electrical resistivity of the material is proportional to the square of the magnitude of the magnetic field when applied at right angles to the direction of the current (Wright, 1966). The resistivity saturates when the field intensity is increased to a critical value and the linear relationship disappears. The effect is most prominent with semiconductors and indium arsenide is a highly suitable material for the construction of a device called a magnetodiode.

A typical magnetodiode is a small thin rectangular plate consisting of an intrinsic semiconductor doped n-type at one end and p-type at the other. The larger intrinsic portion forms a recombination zone where the field is active. Stanley *et al.* (1986) quote typical dimensions of 3.0 mm × 0.6 mm × 0.3 mm for the active zone. The frequency response is flat at frequencies from 0 to 3 kHz and the sensitivity remains constant at room temperatures. Magnetodiodes have greater sensitivity than a Hall probe, but for reasons given earlier, they are not suitable for use with high magnetic fields.

Magnetoresistance also occurs in films of permalloy (81% Ni, 19% Fe) 10–100 nm thick and deposited on smooth substrates.

### 3.5.4 Coil detectors

Coil detectors, until recently the most commonly used flux leakage probes, have the advantages of being cheap, hard-wearing (provided they are properly encapsulated) and readily shaped and sized to the requirements of specified applications. They are particularly suitable for scanning large surface areas. Values of flux density can be determined either by measuring the self-inductance of the coil or, when it scans a surface at a constant speed, the induced EMF (equation 2.18). Coils, however, are generally much larger than semiconductor probes, and complications can arise when they scan across rapidly changing fields, especially when their axes are parallel to the surface, as is necessary to detect the $x$-component of the flux density (section 3.2.3). Fine cracks can easily be missed by coil sensors; this emphasizes the need for earlier magnetic particle inspection wherever possible.

Equation (2.49) shows that the self-inductance $L$ of a coil is proportional to the magnetic permeability $\mu$ of the space enclosed by its windings, i.e. the core, and the magnitude of the flux density $B$ at a point in a leakage field can thus be determined from measurements of $L$. The maximum resolution in determining spatial changes in $B$ is obtained by using coils which are as small as possible, taking into account the corresponding decrease in the values of $L$. Variations of $L$ are measured using an AC bridge; this requires the flow through the coil of a current which, although very small, produces a magnetic field. However, with an air-cored coil having minimal dimensions, the strength of this field is usually small compared with leakage fields, especially at high frequencies. Effects of eddy currents induced by the coil, especially when scanning with its axis vertical, are virtually eliminated by the proper phasing out of any resistive component of the impedance of the coil.

Another type of coil detector is the Förster Microprobe (Stanley *et al.*, 1986) which consists of a small ferrite core, typically 2 mm long and 0.1 mm in diameter, surrounded by a coil. Ferrites possess very high magnetic permeabilities and, being semiconductors, they are virtually free from eddy currents. The coil is excited at a relatively high frequency, e.g. 140 kHz. This excitation is combined with the excitation from the leaked flux and the resultant output is detected through a filter network tuned to double the input frequency. The sensitivity can be increased by using two concentric coils, one transmitting and one receiving.

The difficulties imposed by passing currents through coils are avoided by scanning with an unexcited coil at a constant speed across the surface of the test sample. As the coil cuts through the lines of magnetic flux, an EMF is induced in accordance with Faraday's law (equation 2.18). Greater sensitivity in detection is achieved by using a ferrite-cored coil for the reasons given above. It is important the scanning speed is kept constant because the induced EMF is proportional to the *rate* of change of flux density. The detected EMFs pass through integrating devices to provide evaluations of the flux density $B$.

### 3.5.5 The SQUID

A comparatively recent advance in the measurement of magnetic fields is the introduction of a magnetometer called the superconducting quantum interference device (SQUID). The SQUID operates on the principle that a magnetic field induces an electric current in a closed superconducting circuit. It can detect very weak fields with a degree of precision considerably higher than any other form of magnetic detector. A SQUID can measure magnetic fluxes having values down to $2.07 \times 10^{-15}$ Wb at distances of up to about 90 mm from the object being tested. It is very highly sensitive to localized variations in field strengths but insensitive to strong external fields. These properties help to provide a means of measuring very weak values of magnetic flux linkage resulting from the presence of fine cracks. A high degree of precision and repeatability of measurements results from the fact that a SQUID indicates only quantized values of magnetic flux, i.e. exact multiples of $2.07 \times 10^{-15}$ Wb. This quantity is equal to the ratio $h/e$, where $h$ is Planck's constant, $6.63 \times 10^{-34}$ J s, and $e$ the electronic charge, $1.60 \times 10^{-19}$ C.

An essential characteristic of a SQUID is that any electrical conductor used in its construction has zero resistance when in use, as would be the case if the conductor were at a sufficiently low temperature to be in a superconducting state, i.e. if it became a **low-temperature superconductor** (LTS). For most conventional metals this would occur at temperatures of below 30 K ($-243\,°C$), achievable only by cooling with liquid helium to temperatures below the critical value $T_c = 4.2\,K$ ($-269\,°C$) at standard pressure. The next highest obtainable temperature is 77 K (−196 °C), the boiling point of liquid nitrogen, but far too high to produce superconductivity in metals.

The procurement and handling of liquid helium involve a great deal of expense and careful consideration, hence much effort has been made towards finding materials displaying superconductivity at temperatures above the boiling point of liquid nitrogen. Woods (1996) has drawn attention to the work of Bednorz and Müller (1986) on the discovery of a family of semiconducting materials displaying superconductivity at around 30 K and also to later work by Wu *et al.* (1987), Kaneko *et al.* (1991) and Huang *et al.* (1993) on materials which are superconducting at temperatures within the range 90–135 K (−183 to −138 °C). Each of these materials is known as a **high-temperature superconductor** (HTS). High-temperature superconductors are specially developed compounds with copper as a common constituent. One of the first to be discovered is commonly known as YBCO-123 and has a critical temperature of 90 K, well above the boiling point of liquid nitrogen (77 K). More recently developed materials such as those in the Bi–Sr–Ca–Cu–O and Tl–Ba–Ca–Cu–O systems (Woods, 1996), with critical temperatures at the higher end of the range 90–135 K, appear to be simpler to use than YBCO-123. Liquid nitrogen is very much cheaper and more convenient to use than liquid helium but great care must be taken to keep its temperature below 90 K. However, as explained by Woods, better results can be obtained with low temperature superconductors, partly because of lower noise levels, but this is of greater interest to research physicists than to NDT practitioners.

A short but lucid account of the underlying principles of SQUIDs has been given by Hands (1985), who shows how their operations rely on the **Josephson effect** and the quantization of magnetic flux. This effect may occur between two pieces of superconducting material when joined by an insulating layer of the order of 0.2 μm thick; the junction is called a **Josephson junction**. Pairs of electrons, known as Cooper pairs, are able to tunnel through this layer, consequently producing a superconducting loop. The superconducting loop creates a magnetic flux that is an integral multiple of $2.07 \times 10^{-15}$ Wb.

A longer and easily readable account of the use of SQUIDs has been produced by Clarke (1994). This is highly recommended in spite of the fact that NDT applications are not covered, which probably explains why he somewhat underestimates the amount of interest in the use of RF SQUIDs.

SQUIDs fall into two categories, DC and RF (Figure 3.19a). In each case the circuit consists of a suitably shaped superconducting film (S) of surface dimensions approximately 200 μm × 200 μm; the film is sprayed on to a substrate surface, typically 10 mm square, made from a material such as magnesium oxide. To create the Josephson

# Quantitative flux leakage detectors

junction, the substrate is etched to provide a step (E) having a height equal to the appropriate insulating thickness and bisecting the narrow parts of the SQUID's cross-section at right angles, so as to form the required weak link. Coupling between the divided parts of the film may be achieved using a suitable metal (e.g. silver) thin film. Potential differences will thus appear across the Josephson junctions.

The DC SQUID consists of a loop having two weak links, so as to produce two parallel paths for the current. External connections are required to complete the circuit in order to provide measurements. With the RF SQUID there is only a single weak link and the complete circuit is contained within the SQUID, thus avoiding the need for any connections. This circuit is inductively coupled to a resonant circuit and only the voltage generated in it is required to evaluate the magnetic flux. Figure 3.19b illustrates the corresponding electrical symbols representing the two different types of SQUID.

The DC SQUID is more sensitive because it uses a single screen to surround both it and the source of the magnetic field to be measured. As with biomagnetic applications, it is used for NDT measurements at the minimum levels, RF SQUIDs are less sensitive but adequate for most NDT applications. Even with the noise they encounter, they have considerably greater sensitivities than the more conventional devices

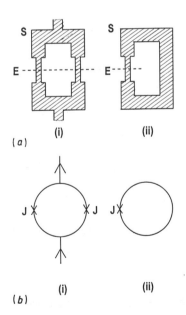

**Figure 3.19** (a) SQUID schematics: (i) DC and (ii) RF. S = the high-temperature superconducting thin films that are coated onto crystalline substrates (not shown); E = etched steps. (b) The corresponding circuit symbols: J = Josephson junction. (Part (a) reprinted from Woods, 1996, by permission of the Institution of Electrical Engineers)

such as Hall probes and induction coils. They do not require electrical contacts, so that their screening requirements are less stringent. It is thus possible to place an RF SQUID at distances of up to 90 mm from the test object.

### 3.5.6 Eddy current detectors

For eddy current testing (section 5.3), a coil is normally used as a receiver, which could be either the transmitting coil in the receiving mode or a separate receiving coil. Here the degree of sensitivity depends on the value of the penetration depth of the material under test (equation 2.90) and there may be limitations on this value. These limitations may be overcome by using a magnetoresistive element to scan the object under test at a constant speed, so as to ensure a suitable rate of change of magnetic flux density produced by the eddy currents (Aurin, 1996). A higher degree of sensitivity can be achieved by using a SQUID (section 3.5.5) as long as it is possible to overcome any difficulties caused by low temperatures.

## 3.6 QUANTITATIVE FLUX LEAKAGE APPLICATIONS

### 3.6.1 General considerations

The relevant equipment for quantitative flux leakage testing may be comparatively sensitive for defect sizing but it is expensive to buy and maintain. Thus, for surface testing on a small scale, alternative methods such as eddy current testing (Chapters 4 and 5) might be considered. However, the methods are highly advantageous for the large-scale automated testing of ferrous steel billets, rods, tubes and wires using rapid scanning. Billets can be tested effectively by magnetography (section 3.4) but rods, tubes and wires require the use of specially designed scanning systems. Some examples of these systems are briefly described below. For additional information, the reader is referred to Stumm (1985).

No British Standards exist at present for quantitative flux leakage methods but the following ASTM standard is available: E 570 Flux leakage examination of ferromagnetic steel tubular products.

### 3.6.2 Testing tubes, rods and plates

An extensive range of devices has been developed by Förster for the automatic flux leakage testing of tubes and rods, having diameters in the range 10–500 mm and greater, with externally located sources and sensors. One of them, called the Rotomat (Figure 3.20), was designed for the detection and sizing of cracks at both the external and internal surfaces of seamless and seam-welded tubes. A direct field is applied in the circumferential direction to achieve a sufficiently high degree of sensitivity in sizing longitudinal defects; the field is applied by an encircling rotating yoke (Figure 3.20a) through which the tube is fed at a constant speed of up to $2\,\mathrm{m\,s^{-1}}$ in an axial direction. A pair of pole pieces provides a magnetic circuit around the tube, which is separated from them by a narrow air gap. The lengths of the pole pieces can be adjusted to accommodate tubes having different diameters. Two diametrically opposite probe heads, each containing an array of typically 16 probes covering a width of 80 mm are attached to the yoke so as to provide a complete scan of the tube.

Figure 3.21 indicates how defects at the outer and inner surfaces can be distinguished from one another. As mentioned in section 3.1, the density of flux leakage is less at inner defects than at outer defects but the leakage takes place over a greater circumferential distance. The output of the voltage, initially direct, is directly proportional to the rate of change of flux during scanning at a constant speed and it becomes alternating as the detector passes across a region of flux leakage. The frequency of the output is thus higher for external cracks than for internal cracks. The position of a crack can then be indicated with the aid of a suitably designed filter. Crack depths of the order of 0.4 mm at both external and internal surfaces can be measured by this method. For tubes with welded seams, the weld area is identified by a change in magnetic permeability which in turn gives rise to a change in sensitivity for crack detection.

An instrument which can be used in a complementary manner to the Rotomat is Förster's Transomat, where the direct energizing field is induced in an axial direction by a pair of coils (Figure 3.22) so as to facilitate the detection of transverse defects. The heads of the probe array are arranged in an annular manner to give complete diameter coverage, and their position can be adjusted to accommodate tubes having different diameters. Discrimination between defects at external and internal surfaces is achieved in the same way as with the Rotomat. A complete scan for defects in any direction can be achieved by passing the tube through the two instruments in succession.

A higher sensitivity for the detection of defects at the external surface of a steel tube or rod can be obtained by applying an alternating magnetizing field having a sufficiently high intensity to produce saturation in the region of the surface (section 3.3.3). Because of the attenuation of the magnetic field with depth, the saturation vanishes below the surface skin, so as to produce an increase in magnetic permeability, hence flux density. This has the effect of decreasing the thickness of the 'magnetized' zone and apparently

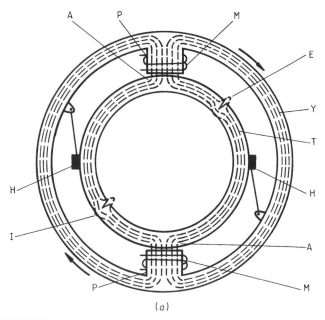

**Figure 3.20** (a) The magnetizing yoke for Förster's Rotomat: A = air gap, E = flux leakage from external defect, H = probe head, I = flux leakage from internal defect, M = magnetizing winding, P = pole piece, T = tube, Y = yoke ring. (b) Using the Rotomat to test a tube.(Part (b) reprinted by permission of the Institut Dr Förster)

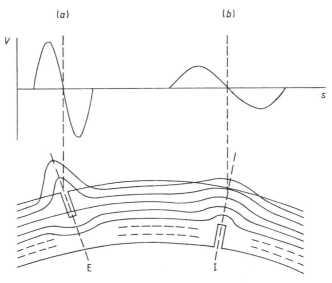

**Figure 3.21** Variations of probe output voltage $V$ with displacement $s$ for the circumference of a rotating tube tested with the Rotomat: (a) external defects (E) and (b) internal defects (I). (Reprinted by permission of the Institut Dr Förster)

increasing the thickness of any cracks, as indicated by the flux leaking through the saturated region (Figure 3.23).

An instrument which applies this principle is Förster's Circoflux (Figures 3.24) which generates a field at a kilohertz frequency with a power of several kilowatts. The use of high frequencies means it is possible to suppress the low-frequency noise originating from mechanical devices such as scanners. Furthermore, the resulting decrease in skin depth ensures a higher resolution of detection. Any residual magnetism after testing is minimal. A pair of yokes and probe

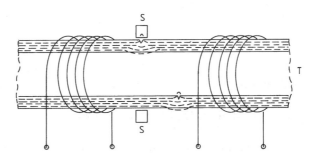

**Figure 3.22** Side elevation of tube (T) tested by the Transomat: the diagram shows two probes (S) which form part of a circular array. (Reprinted by permission of the Institut Dr Förster).

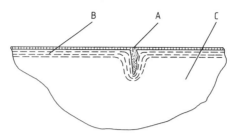

**Figure 3.23** High-intensity magnetization of a test sample containing a crack (A). Magnetized using an alternating field, the sample exhibits a saturated region (B) and a high-permeability region (C). (Reprinted by permission of the Institut Dr Förster)

arrays are mounted in a test head which rotates as the rod or tube passes through it at a constant speed of up to $2\,\mathrm{m\,s^{-1}}$. The equipment is designed to test rods and tubes having diameters of 25–260 mm.

For tubes such as buried pipelines, the external surfaces are inaccessible and the magnetizing yoke and probe arrays are mounted on a carriage which is propelled through the inside of the tube. The application of direct magnetic fields is necessary to allow sufficient penetration for detection and the gaps between the pole pieces of the yokes and the tube wall should be kept to a minimum. Furthermore, precautions should be taken to prevent the carriage from wobbling. Although the possibility of the formation of cracks caused by stresses exerted by the medium surrounding the pipe cannot be disregarded, the most likely defects are those resulting from corrosion and its associated stresses, which can occur at both external and internal surfaces (Atherton, 1983).

When relatively short lengths of tube are tested, e.g. of up to a few hundred metres, the main items of instrumentation can be sited externally and connected by a cable to the carriage, which can be drawn by means of a rope or other device at a constant speed. However, when testing long tubes such as pipelines having unbroken lengths of several kilometres, the assembly is contained in a vehicle which also carries any necessary instrumentation, a portable electric power supply and a drive unit. With field excitation by battery, difficulties may arise in maintaining a sufficiently high magnetizing field for any long period.

An inspection system capable of locating defects in long pipelines with the magnetic flux leakage method has been developed by British Gas plc and is called an 'intelligent pig' (Figure 3.25). The carriage, or pig, is propelled through the pipe by the flow of the fluid under pressure (e.g. aviation fuel, gas, oil); besides the magnetizing unit and probes, the pig carries a specially developed data processing unit, tape

## Quantitative flux leakage applications

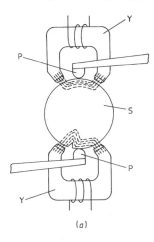

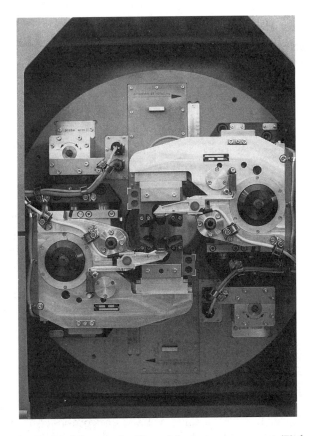

**Figure 3.24** (a) Yoke (Y), sample (S) and probe arrangement (P) for the Circoflux. (b) Rotating heads of the Circoflux. (Reprinted by permission of the Institut Dr Förster)

**Figure 3.25** Intelligent pig: an inspection carriage for testing gas pipelines by magnetic flux leakage. (Reprinted by permission of British Gas plc)

recorder and power pack housed in impact-resistant cylindrical modules. The pigs are made in different sizes to fit pipes having diameters of 200–1200 mm. Corrosion defects can be located to within 1.5 m.

More recent work with the development of intelligent pigs for the internal testing of ferromagnetic tubes has been described by Jansen and Festen (1995) and by Jansen *et al.* (1994), who investigated the various sources of error giving rise to either oversizing or undersizing of defects. An electromagnet and a Hall probe, rigidly connected to one another by mechanical means, minimized or eliminated any defects by ensuring a proper design of the magnetic system in the pig, thereby obtaining a sufficiently high degree of magnetization to allow penetration to the outer surface of the wall. As a result of converse magnetostriction (section 3.8.4) the value of the magnetic flux density also depends on the pressure in the pipe; this needs to be taken into account.

The testing of ferromagnetic plates and other objects having plane surfaces presents few problems. The basic equipment consists of a Hall probe mechanically coupled to a permanent magnet. The magnet and probe combination is then used to scan the surface of the object under test, as described by Stanley (1996), who also discusses devices for testing ferromagnetic tubing. An important application of this method is the inspection of floors in above-ground storage tanks (e.g. Johnston, 1992).

### 3.6.3 Testing steel wire ropes

Steel wire ropes are extensively used for mine hoists, passenger lifts and cable cars and may be up to several kilometres long. Because any failure can be catastrophic, regular inspections are necessary. The flux leakage method has proved to be highly effective for testing these ropes in view of its reliability, simplicity and speed, especially if a complementary eddy current test is made. Tests are conducted by scanning with a unit containing a magnetic field generator and a suitably designed probe head.

The relevant steel wire rope consists of several strands, typically six, wound in a helical manner about a straight strand called the core. Each strand may consist of about 30 wires wound in a similar manner. Bergander (1985) has enumerated the various defects arising in service, which include abrasion, reduction in diameter due to various causes, kinking, heat damage, peening (i.e. permanent distortion caused by plastic flow), fatigue failure and breakage of individual wires. Many defects can be tolerated for some time, and a rope is only replaced at a stage when the defects increase in number at an unacceptable rate. An insidious defect is internal corrosion caused by leakage of water into the rope.

Both alternating and direct field excitations can be used for rope testing; alternating fields provide greater sensitivity to generalized corrosion and wear, direct fields provide greater sensitivity to broken wires and internal corrosion (Poffenroth, 1985). The AC method is effectively an eddy current method which detects changes in magnetic permeability as well as changes in electrical conductivity. It uses two coils coaxial with the wire; a primary excites the field at a low frequency (either 10 or 30 Hz) and a secondary acts as a receiver, changing its impedance as a result of any detected flux leakage. The field is oriented in the axial direction of the wire. For the DC method, a permanent magnet induces an axial field in the wire. A search coil is located between the poles of the magnet in such a way as to pick up any radial component of flux leakage. With both methods the scan is made at a constant speed.

Recent work on the testing of steel wire rope has been reported by Weischedel and Chaplin (1991), Geller *et al.* (1992) and Hanasaki and Tsukada (1995).

## 3.7 MAGNETIZATION AND HYSTERESIS METHODS

### 3.7.1 General considerations

Magnetization and hysteresis measurements (section 2.7.1) have been extensively used for the non-destructive testing of components made from ferromagnetic metals to determine quantities such as coercivity (or coercive force), retentivity, saturation field and maximum permeability. These quantities can then be applied to assess factors such as hardness, chemical composition, degree of internal stress, impurity concentration and grain size. They can also be used to detect defects and to measure sizes of objects. Two different techniques are used for hysteresis measurement, one where the conventional $B/H$ hysteresis loops are obtained and the other, developed by Förster and called the Magnatest, where $B$ is recorded as a function of time, usually for a complete cycle. Encircling coils are mostly used for field excitation, although surface-scanning coils are usable for surface inspection (section 4.1). The frequencies employed normally vary from just above zero to mains frequency (50 to 60 Hz), but the upper limit is 120 kHz with the Magnatest. The choice of frequency depends on the required degree of penetration and whether or not eddy currents need to be suppressed. Objects of any shape can be tested and the upper limit to size is restricted only by the practicability of constructing a large enough coil. The component to be tested should be protected from any unwanted magnetic fields.

### 3.7.2 $B/H$ loop method

With the simplest form of hysteresis testing of a cylindrical rod or tube, the magnetic field of magnitude $H$ is excited by AC at say 50 or 60 Hz through an encircling coil and the magnitude of the resultant flux density $B$ is detected by a probe placed in a suitable position near the object being tested. The potential difference $V_x$ across a resistor having a suitable value and connected in series with the exciting circuit is fed through an amplifier to the $x$-plates of a cathode ray oscilloscope. The potential difference $V_x$ is proportional to the exciting current and, hence, to $H$. The potential difference $V_y$ generated by the probe, and proportional to the rate of change of magnetic flux (equation 2.18), is fed through an integrating device then an amplifier to the $y$-plates of the oscilloscope so as to indicate $\int V_y \, dt$, which is proportional to $B$. Graduations of the oscilloscope graticule for $H$ and $B$ are obtained from a knowledge of the relationship between the current and magnetic field for the exciting coil and by knowing that, in the absence of a sample, $B = \mu_0 H$ where $\mu_0 = 4\pi \times 10^{-7}$ H m$^{-1}$. For very low frequency measurements, a clearer display can be obtained with a storage oscilloscope.

Förster has shown experimentally that the effect of eddy currents is to increase the area of the hysteresis loop without altering its shape. Förster had two ring samples cut from the same rod of steel, i.e. with the same external radius, but with one of them (b) having a wall thickness $\sqrt{2}$ greater than the other (a); he obtained hysteresis curves for each of them in turn at frequencies of 0 and 60 Hz. The area $A_a$ of the hysteresis loop for sample (a) was assumed to be given by

$$A_a = H_1 + E_1$$

where $H_1$ is the purely magnetic contribution obtained at zero frequency, and $E_1$ the contribution due to eddy currents at the test frequency. Förster found that the area of the loop for sample (b) was larger than for sample (a) by the amount $E_1$ at the 60 Hz frequency:

$$A_b = H_1 + 2E_1$$

from which

$$A_b - A_a = E_1$$

so that an increase in tube wall thickness by a factor of $\sqrt{2}$ gives rise to a doubling of additional area of the curve due to the induction of eddy currents, i.e. electrical conductivity effects.

### 3.7.3 Barkhausen effect

The Barkhausen effect was discovered early in the twentieth century but it had very little application to NDT until the late 1970s; rapid progress has been made since then. The reader interested in the recent development of the Barkhausen effect and its sister phenomenon of acoustic emission is recommended to read the introduction to a paper by Hill et al. (1993).

When a ferromagnetic material is magnetized, the increase in flux density takes place in a discontinuous manner as a result of the magnetic domain walls rotating and breaking away progressively from the pinned-down positions occupied in the initially unmagnetized state. This phenomenon is called the **Barkhausen effect** (section 2.7.1) and it becomes prominent when the value of the magnetizing field is of the order of the coercivity. Each breakaway provides a source of acoustic emission. The accompanying electromagnetic emission is called either Barkhausen emission or Barkhausen noise, but Barkhausen noise may be misleading because 'noise' is normally defined as an undesirable signal; hardly the case here.

The breaking away is affected by the presence and density of impurities and dislocations in the crystalline structure and by any grain boundaries; the manner in which it takes place is affected by both external and internal stresses. The discontinuous nature of the effect allows it to be measured by a pulse-counting device, using either the output of a magnetic flux probe or an ultrasonic transducer that picks up the accompanying acoustic emission.

A plot of the count rate, i.e. $dn/dH$ against $H$ (Figure 3.26a) takes the form of one or more peaks with different sizes. The values of $H$ at the peak centres and the nature of the peak structures can be used to identify and assess the extents of the various mechanisms contributing to the Barkhausen effect for the sample being tested. A review of this and other magnetic methods of testing has been given by Jiles (1988), including a comprehensive bibliography.

Advantage may be taken of the fact that Barkhausen emission has both acoustic and electromagnetic components. Because of the skin effect (equation 2.90), surface and subsurface phenomena may be investigated using the electromagnetic component; however, the acoustic component can be used for testing bulk properties. This has been clearly shown by Hill *et al.* (1991) in their determinations of hardness and grain size in nickel. Work in these fields by Hill *et al.* (1993) has been extended with success to testing iron and steel welds.

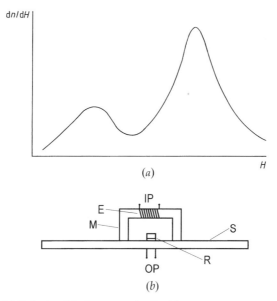

**Figure 3.26** (a) Relationship between the Barkhausen count rate $dn/dH$ and the magnetic field strength $H$ for two different defect distributions. (b) Arrangement for observing Barkhausen emissions from a sample undergoing hysteresis: E = exciting coil, IP = input, M = magnet, O = output, R = receiving coil, S = steel sample. (Part (a) reprinted, by permission, from Jiles, 1988, copyright Butterworth-Heinemann Ltd; part (b) reprinted from Dhar *et al.*, 1992, by permission of the American Society for Nondestructive Testing)

Dhar *et al.* (1992) used the Barkhausen effect to determine easy axes of magnetization in steels. These axes are in the direction of preferred orientation, perhaps created by rolling, drawing or annealing, i.e. the material is rendered anisotropic. Figure 3.26b shows an arrangement for observing the effect by submitting a pipe-steel sample to a hysteresis cycle. The signal picked up by the receiving coil is pre-amplified then passed through a bandpass filter so that the Barkhausen effect can be detected. The output signal varies in amplitude as the pipe is rotated about its axis, thus indicating the degree of anisotropy.

At present, no standards have been laid down regarding methods of exciting magnetic fields, choice of detection frequency and excitation of waveform. These matters have been studied by Sipahi *et al.* (1993), whose main conclusions are that the Barkhausen effect can be distinguished on the basis of frequency content. Both sinusoidal and triangular waveforms are suggested, but it may be better to vary the magnetic field with time so the corresponding variation of flux density is constant. Difficulties arise with square waves as a result of their high-frequency content. Significant improvements in resolution when measuring Barkhausen emission have been achieved by making the receiving coil from a recording head ('read head') taken from a $3\frac{1}{2}$ in (8.75 mm) floppy disk and mounting it in a U-shaped laminated sweep field core (Krause and Atherton, 1994). Slit widths of 0.25–0.7 mm could be detected in a steel sample.

### 3.7.4 Coercivity measurement

If the size of a test sample is of no concern but only its overall physical properties, hardness, degree of heat treatment, etc., it may be sufficient to measure only coercivity. This can be done by magnetizing the sample to saturation by applying a constant magnetic field; when the constant magnetic field is slowly reduced to zero there is a residual flux density, i.e. retentivity *OC* (Figure 2.13). A reverse field is then applied until the retentivity disappears; the value of this field is equal to the coercivity *DO* (Figure 2.13).

An instrument designed to perform this function is Förster's Koerzimat. The sample is placed in a non-metallic holder and inserted well inside a solenoid coil screened from unwanted magnetic fields. The operation can be carried out either manually or automatically and the value of the coercivity is displayed digitally.

### 3.7.5 Incremental permeability method

A device called the CeNteSt 91 ferromagnetic tube testing equipment (Owston, 1985) has proved successful for the internal examination of defects such as pits, holes, corrosion and erosion in the walls of ferritic steel tubes of the type used for boilers, refrigeration plant and heat exchangers, with outside diameters of 16–90 mm and wall thicknesses of 0.5–10 mm. The allowable wall thickness depends, to some extent, on the diameter of the tube.

The probe (Figure 3.27a) is connected to the main body of the instrument by a cable of suitable length. Scanning is performed by threading the probe through the tube (Figure 3.27b). The size of the probe should be selected to allow a reasonable fit in the bore of the tube. Probes having an adjustable diameter are available for tubes of

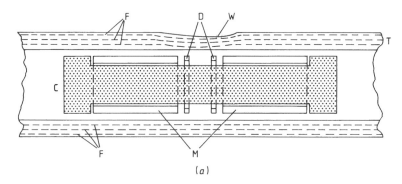

(a)

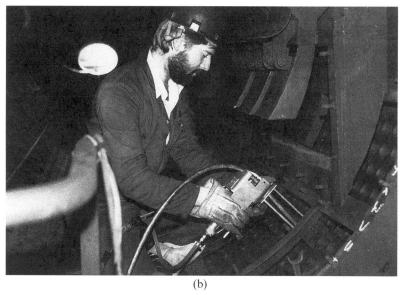

(b)

**Figure 3.27** CeNteSt 91 probe inside a steel tube (T) with wall thinning due to corrosion and showing lines of magnetic flux: C = core of DC electromagnet, M = magnetizing coil, D = detecting coil, W = wall thinning, F = magnetic flux lines. (b) Using a CeNteSt 91 probe to inspect steel boiler tubing. (Part (a) reprinted from Owston, 1985, by permission of the British Institute of Non-Destructive Testing; part (b) reprinted by permission of CNS Electronics Limited)

larger bore. The presence of deposits, up to 1 mm thick, in the bore does not usually have any effect on the performance of the instrument.

The probe consists of an electromagnet, energized by DC, and a pair of AC-excited coils connected differentially (section 5.4.3) and tapped at the point of connection to enable the inductance of either coil to be measured. The value of the DC should be sufficient to give a field producing a high level of magnetization within the tube wall but to remain well below saturation. The differential coils carry a low-amplitude AC at a frequency of the order of 25 kHz but only the region of the tube characterized by the degree of penetration $\delta$ (equation 2.90) from the inside surface is affected. The resultant field variation in this region gives rise to a minor hysteresis loop (e.g. *PQ* in Figure 2.13).

The presence of any defect causes the lines of magnetic flux induced by the electromagnet to converge, hence they provide increases in the values of the flux density $B$ in all regions, including just below the surface. This phenomenon has the effect of shifting the origin of the minor hysteresis loop, hence it changes the value of the incremental permeability $\mu_{inc}$. Because the values of the inductances of the detecting coils depend on $\mu_{inc}$ (equation 2.49), the changes in their impedances are characteristic of the defect. The coil detectors are connected differentially in order that only variations of impedance arising from the presence of small defects such as corrosion pits are detected, thus giving an increased sensitivity of detection (cf. eddy current differential coils, section 5.4.3). For the assessment of longer defects such as wall thinning, which equally affects both coils in the differential arrangement, the change of inductance of only one of the detecting coils is measured. The relationship between the output indications of the instrument and the extent of the defect, e.g. wall-thickness changes or sizes of pitting holes, can be obtained by calibration using models containing simulated defects.

The method has the advantage that it can detect defects anywhere within the tube wall, including the outer and inner surfaces, not just within the skin depth, corresponding to a frequency of 25 kHz. However, although it is capable of detecting circumferential cracks, it cannot detect longitudinal cracks unless they appear at the inner surface of the tube wall.

More recent work on the incremental permeability method has been carried out on the testing of boiler tubes by Grimberg *et al.* (1996).

### 3.7.6 Magnatest method

The Magnatest series of instruments was designed by Förster to test ferromagnetic materials by subjecting them to hysteresis and observing the change of magnetic flux with time over a single cycle. When operated at frequencies above a few hertz, there is sufficient induction of eddy currents to provide information about the effects of the electrical conductivity and magnetic permeability of the tested object. Using encircling coils, this method also provides a means of sizing small objects of regular shape, such as ball-bearings.

With the original form of the Magnatest equipment (McMaster, 1963e), the properties of the test sample were compared with those of a standard, and an analogue display appeared on the screen of a cathode-ray oscilloscope. This has now been superseded by a computer-aided instrument called the Magnatest S in which the relevant properties of standard samples are stored for reference. This equipment has a visual display unit and uses a single transformer-type coil, either encircling or surface probe; the sample to be tested is located in a fixed position relative to the coil so as to maintain constant testing conditions. For example, a steel rod under test can be located in the manner indicated in Figure 3.28 to ensure it is coaxial with the encircling transformer coil.

An alternating current (which can be varied by up to a maximum of 2 A) is applied at a frequency of 2–128 kHz to the external primary winding of the coil. The magnitude $B$ of the magnetic flux density is sensed by the secondary coil and the instrument records its variation with time. For low values of applied field, the hysteresis curve is virtually linear and the variation of $B$ takes place at the field frequency. For higher magnetic fields, the curve is no longer linear and harmonics appear in the oscillations of $B$. The contents of the resulting spectrum depend on the shape and size of the hysteresis curve, hence on the properties of the test object as well as on the magnitude and frequency of the applied field. The instrument can evaluate the signal amplitudes at frequencies of up to the seventh harmonic.

This instrument can thus be used to test components made from ferromagnetic metals for properties such as overall hardness, depth of

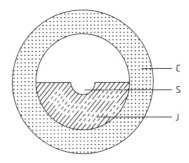

**Figure 3.28** Using a jig (J) to locate a rod or tube concentric with a Magnatest coil (C). The jig is made from a non-ferromagnetic dielectric material, e.g. wood or plastic, and the semicylindrical cavity (S) provides a close fit for the test object.

case hardening, composition, degree of heat treatment, the presence of internal stresses, machinability, dimensions, etc. To perform any of these tasks a calibration is made with standard samples of a given material and having properties of known values. For example, to calibrate for surface hardness, say 10 steel rods of identical size and material properties but different known hardnesses are sequentially inserted into the coil, the corresponding numerical values are fed into the computer and a regression analysis is performed. The computer then selects the optimum values of magnetic field and frequency to find a parameter having a distribution as close as possible to that for hardness, e.g. the signal strength of the third harmonic for a current of 1.5 A and a fundamental frequency of 32 Hz. This information is stored for use whenever a hardness test for this particular type of sample is required, for which a read-out of the value of hardness together with a standard deviation is provided.

## 3.8 LESS COMMON MAGNETIC TECHNIQUES

### 3.8.1 The magnetic balance

The magnetic balance utilizes the force of attraction existing between a magnet and a ferromagnetic object. In its common form it consists of a bar magnet attached to a spring, which can be either compressed or stretched according to the design of the apparatus. The tensile or compressive force on the spring is equal to the magnetic force of attraction. This type of instrument has proved satisfactory for measuring thicknesses of non-ferromagnetic coatings such as paints, plastics and silver plating on ferrous bases.

A typical device of this sort can determine coating thicknesses of up to 0.5 mm to a precision of 10% (Blitz *et al*. 1969b) (Figure 3.29). The magnet, which is graduated along its length, is suspended by a spring from the top end of a pencil-shaped non-ferromagnetic tube about 150 mm long, and when the instrument is not in use, the lower ends of the magnet and tube are at the same level. When the device is placed vertically on the surface under test then slowly raised, the magnet remains in contact with the surface until the tension in the spring is sufficient to overcome the force of attraction. The number of graduations on the magnet which are exposed when the tube is raised can be directly related to the coating thickness, if a calibration has been made previously for coatings of known thicknesses. A more precise and sophisticated, but considerably more expensive, coating thickness detector is the eddy current device described in section 5.5.3.

An instrument with a higher degree of precision and which utilizes an electronically operated extensometer for measuring the attractive

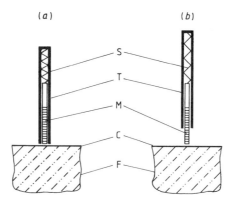

**Figure 3.29** A magnetic spring-balance measures coating thicknesses on ferromagnetic substrates: (a) magnet M in contact with surface coating C; (b) magnet M remote from surface coating C. Other symbols: S = spring, T = non-ferromagnetic tube, F = ferromagnetic substrate.

forces has been applied to monitor corrosion in aircraft turbine blades made from Inconel and other high-grade ferromagnetic alloys by scanning the surface (Pizzi and Walther, 1979). The instrument is designed to keep the magnet at a fixed distance from the surface and the force of attraction is directly related to the magnetic susceptibility, i.e. $\mu_r - 1$, of the material under test. The magnetic susceptibility changes when corrosion takes place because of the resultant changes in the structure of the material.

### 3.8.2 Magnetic reluctance measurements

Measurements of reluctance (section 2.7.2) can be carried out by placing a magnetic yoke in contact with the surface of a ferromagnetic material under test to form a magnetic circuit. The magnetic flux density is measured at some convenient point in the circuit usually on the yoke.

Guehring (1982) describes how the method can be used to determine thicknesses of nickel coatings on non-ferrous substrates. The yoke is excited with a direct field and the flux density measured with low-frequency AC coils. The reluctance of the circuit is decreased as the thickness of the coating increases. Care should be taken that the applied field is well below saturation level for the coating material, which is considerably less for nickel than for iron.

Emerson (1976) describes a method for measuring case-hardening depths in steels; this phenomenon affects the value of magnetic permeability, hence the reluctance. A Hall probe was used for measuring the flux density in a DC-excited yoke. Difficulties arose because of reluctances being introduced by uncertain contacts between the yoke and the surface of the test object. The sensitivity of detection was not as high as experienced with coercivity measurements.

### 3.8.3 Magnetic debris testing

A problem arising with the operation of oil-lubricated machinery is the accumulation of wear debris. This should be regularly cleared to avoid the risk of damage. In most cases the machinery is of steel construction and the debris is thus ferromagnetic by nature. The principal methods employed are spectrometric oil analysis, ferrography and magnetic plug inspection. The first method is not a magnetic method; it is used to identify minute debris particles up to about 10 μm in size.

Two types of ferrography, analytical and direct reading, can deal with particles of size up to 50 μm. With analytical ferrography, a sample of oil containing debris is deposited on a slide placed over a magnet. The slide is then removed and examined under a microscope for a sample count and for sizing and identification of the cause of wear. With direct-reading ferrography, the debris is collected in a glass tube and the optical density measured.

Magnetic plugs provide continuous debris monitoring while the machinery is running. They are placed in suitable positions in the oil flow and removed at regular intervals; the debris is carefully transferred to a strip of Sellotape which is then examined under a microscope. Plugs are particularly effective for the examination of large-sized debris particles known as chips.

An account of ferrography and magnetic plug inspection, as applied to the monitoring of roller bearings, has been given by Cooper (1983). Brenneke (1989) has described the use of magnetic plugs to predict early failure in gas turbine aero-engines.

Lees *et al.* (1978) detected accumulations of magnetic oxide debris in the lower bends of austenitic stainless steel tubes (non-ferromagnetic) in electric power station boilers. Their instrument consisted of a permanent magnet located centrally between a pair of probes in a vertical position and housed in a casing. A magnetic field was induced in the tube by placing the instrument beneath it and the resultant flux could be sensed by the probes. In this way they were able to measure the depth of the accumulated ferromagnetic debris.

### 3.8.4 Magnetostriction

When a magnetic field is applied to an object made from ferromagnetic material, the object experiences a mechanical strain in the direction of the field; the strain is either positive or negative, depending on the nature of the material. This is called the magnetostrictive effect. Conversely, when such a ferromagnetic material is subjected to a mechanical stress, there is a change in flux density, either positive or negative, in the direction of the applied stress. This phenomenon can be used to determine mechanical stresses in magnetized ferromagnetic objects from flux density measurements, e.g. with a Hall probe.

Langman (1981) has investigated this phenomenon in loaded mild steel bars with some degree of success, but it does appear that more satisfactory results are attainable using purely mechanical methods.

Robertson (1994) has reported work carried out on specimens of some different types of steels placed in weak magnetic fields, having values comparable to the earth's field and subjected to compressive stresses of up to 200 MPa. Increases in magnetization of the samples, as indicated by increases in values of differential magnetic permeability, were obtained (a) with stress cycling at constant applied field and (b) field cycling at constant applied stress. It was found that steels having similar microstructures displayed similar **magnetomechanical behaviour**.

Kwun and Holt (1995) have studied the testing of lagged steel pipes using a pair of similar magnetostrictive transducers in the form of encircling coils; the coils were separated by fixed short distances and polarization was provided by bar magnets (e.g. Blitz and Simpson, 1996). One of the coils acts as a transmitter, inducing in the tube an ultrasound signal which is received by the other coil. Any defects in the tube wall are indicated by the nature of the received signal.

### 3.8.5 Nuclear magnetic resonance

Nuclear magnetic resonance (NMR) may arise when an object having atomic nuclei with free charges is subjected to direct magnetic fields strong enough to cause nuclei to spin. The value of the spinning frequency depends on the physical and chemical properties of the substance and on the strength of the magnetic field, which should be of constant value. On directing radio frequency (RF) electromagnetic waves to the object, resonance takes place when the wave frequency is equal to the spin frequency, i.e. nuclear magnetic resonance occurs, and a decrease in the amplitude of the RF waves can be observed. Nuclear magnetic resonance has been in use for some time to obtain information about material structures, so it is not surprising that NMR techniques are now being used for non-destructive testing, especially because they do not emit any hazardous ionizing radiations.

Present applications (Smith, 1994) include the monitoring of foodstuff quality, examination of porous media, testing the heterogeneity of polymers and on-site inspection such as the examination of mail and baggage for illegal substances. Saadatmanesh *et al.* (1995) and Saadatmanesh and Ehsani (1997) have used the technique for evaluating the properties of concrete and wood. Other work in this field has been reported by Zick (1994), Strange (1994), Roberts *et al.* (1994), Clayden and Jackson (1994) and de los Santos (1994).

# CHAPTER 4

# Eddy current principles

## 4.1 INTRODUCTION

The eddy current method of non-destructive testing, as currently practised, was pioneered by Friedrich Förster in the 1940s, and there has since been rapid progress in its development. Eddy current tests can be made on all materials which are electrically conducting. They include the sizing of surface and subsurface cracks, thickness measurements for metallic plates and non-metallic coatings on metal substrates, assessment of corrosion and measurements of electrical conductivities and magnetic permeabilities. Electrical conductivities and magnetic permeabilities may be related to structural features such as hardness, chemical composition, grain size and material strength. An important advantage of eddy current testing over some other methods, such as ultrasonic, magnetic particle and potential drop techniques, is there is no need for physical contact with the surface of the object being tested. Thus, careful surface preparation is unnecessary (other than the removal of metallic adherents).

The practice of eddy current testing, discussed in some detail in Chapter 5, consists of exciting an alternating current at a given frequency through a coil, often called a probe coil or simply a probe, located as near as possible to the electrically conducting object being tested, and thus to induce eddy currents in the latter. As a result, changes take place in the components of the coil's impedance, which can be related to the design of the coil, the size, shape and position of the test object and the values of its magnetic permeability $\mu$ and electrical conductivity $\sigma$ of the latter. The impedance of the coil is also affected by localized variations in $\mu$, $\sigma$ and the geometry of the object under test as a result of the presence of a defect. A knowledge of those relationships which are relevant to a given test is necessary for calibration, and this chapter discusses how to derive them.

With the eddy current method, the current passing through the coil generates electromagnetic waves in a nearby electrical conductor (section 2.8.2). The associated magnetic field $H$ induces the flow of

## Introduction

electric currents, i.e. eddy currents, which follow circular paths in planes perpendicular to the direction of $H$, in accordance with equation (2.64*):

$$\text{curl } H = J \qquad (4.1)$$

where $J$ is the current density. The coil and metal sample respectively form the primary and secondary components of a transformer, and the impedance of the coil is consequently affected by the behaviour of the eddy currents. Neglecting its ohmic resistance (i.e. the resistance as measured with DC), this impedance, at a frequency $\omega/2\pi$, is purely inductive and has a value $Z_0 = j\omega L_0$, when completely removed from the test object and any other electrical conducting and ferromagnetic materials. When the coil is located in the testing position, the value of the impedance changes at the same frequency, to $Z = R + j\omega L$. Here the inductance changes from $L_0$ to $L$ and a resistive component $R$ is introduced. The changes in the impedance components are determined by the speed and attenuation coefficient of electromagnetic waves in the test sample (equations 2.88 and 2.89). The corresponding phase changes in the relevant electrical and magnetic vectors, i.e. $B$, $H$ and $J$ are responsible for introducing the component $R$ of the impedance.

In general there are three possible coil arrangements (Figure 4.1), i.e. encircling for tubes and rods, internal axial for tubes, and surface-scanning. Surface-scanning coils are used for testing surfaces which are either plane or have very small curvature within the region directly below the cross-section of the coil. Transformer probes with separate transmitting and receiving coil windings may be used for enhanced sensitivity, perhaps if deep penetration of eddy currents is required in the material under test. The analyses in this chapter are equally applicable to single-coil probes and transformer probes. With transformer probes, the relevant impedance is the impedance of the secondary coil, i.e. the detector.

Before any eddy current test, a calibration should be made with a standard sample. If it is free from any defects, the relationships between the impedance of the components of the coil, the frequency and the electrical and magnetic properties of the material can often be obtained theoretically from one of the analytical methods described in sections 4.2 to 4.5. These relationships are usually approximate because of the need to simplify the analyses, e.g. infinite lengths of encircling coil and sample.

With flaw detection it is the normal practice to calibrate the detecting instruments using test blocks containing modelled defects having various depths. They usually take the forms of saw-cuts to simulate cracks and cylindrical holes to represent voids, but their relationships to real defects are only approximate. Much work has

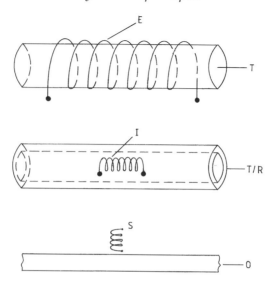

**Figure 4.1** Basic coil positions for eddy current testing: E = encirling coil, I = internal axial coil, S = surface-scanning coil, T = tube, T/R = tube or rod, O = surface-scanned object.

been done, and is still being done, in the use of numerical methods to relate impedance components to positions and sizes of more realistic defects for different designs of probe (section 4.6.3).

## 4.2 COILS ENCIRCLING DEFECT-FREE METAL RODS

### 4.2.1 General considerations

The impedance analysis for eddy currents induced by coils encircling cylindrical conductors was introduced by Förster (1986). The analysis is simplified by considering an infinitely long air-cored solenoid coil of effective radius $r_0$ encircling and coaxial with an infinitely long electrically conducting cylindrical rod having a radius $b$, uniform dimensions and a homogeneous structure, so the directions of the magnetic fields and flux densities are coaxial with the coil.

An alternating current, which varies sinusoidally with time (i.e. as $\exp j\omega t$, where $j = \sqrt{-1}$ induces an alternating magnetic field $H$ having the same frequency and, in the absence of the rod, the resulting magnetic flux density $B_0$ inside the coil is given, at a time $t$, by

$$B_0 = \mu_0 H \qquad (4.2)$$

where $\mu_0 \ (= 4\pi \times 10^{-7} \text{ H m}^{-1})$ is the magnetic permeability of free space and effectively of any non-ferromagnetic material. Both $B_0$ and

## Coils encircling defect-free metal rods

$H$ can be considered to be constant at all points within the coil for given values of excitation current and frequency, because the attenuation of electromagnetic waves in air is virtually zero at the frequencies normally used for eddy current testing (i.e. up to 10 MHz) with the very short distances involved.

With the rod placed in position inside the coil and the values of excitation current and frequency unchanged, let $B = \mu H$ be the magnetic flux density at a given point in that rod. Here $\mu$ ($= \mu_r \mu_0$) and $\mu_r$ are the absolute and relative magnetic permeabilities, respectively, for the material of the rod. $B$ now varies as follows (cf. equation 2.84):

$$\nabla^2 B = \mu_r \mu_0 \sigma \partial B / \partial t \quad (4.3)$$

Since $B$ varies as $\exp(j\omega t)$, for all values of time $t$, we have

$$\nabla^2 B = j\omega \mu_r \mu_0 \sigma B = -k^2 B \quad (4.4)$$

where $k^2 = -j|k|^2$ and $|k| = (\omega \mu_r \mu_0 \sigma)^{1/2}$. Here $\omega$ is the angular frequency, $\sigma$ the electrical conductivity and $j = \sqrt{-1}$. The term $k$ as used here should not be confused with the same symbol which represents wavenumber (section 2.8). Equation (4.4) is expressed in cylindrical coordinates $(r, \phi, z)$ as follows

$$(1/r)\partial/\partial r(r\partial B/\partial r) + (1/r^2)\partial^2 B/\partial \phi^2 + \partial^2 B/\partial z^2 + k^2 B = 0 \quad (4.5)$$

the z-axis is defined here as the common axis of the rod and coil so that $(r, \phi)$ are the polar coordinates in the Cartesian $x-y$ plane. Because $B$ is constant for all values of $z$ and conditions of circular symmetry exist in the $x-y$ plane, $\partial^2 B/\partial z^2$ and $\partial^2 B/\partial \phi^2$ are both equal to zero. $B$ is thus a function of $r$ only, so the use of the vector notation can now be discontinued. Equation (4.5) then becomes

$$\partial^2 B/\partial r^2 + (1/r)\partial B/\partial r + k^2 B = 0 \quad (4.6)$$

This is a modified form of the zeroth-order Bessel equation, the solution of which is obtained from equation (C.2) given in Appendix C. Thus, on applying the boundary condition that $B = \mu_r \mu_0 H = \mu_r B_0$ when $r = b$, the radius of the rod, we have

$$B = \mu_r B_0 J_0(kr)/J_0(kb) \quad (4.7)$$

assuming that $B$ is finite for all values of $r$, where $J_0$ indicates a Bessel function of zeroth order.

The magnetic flux $\Phi$ through the cross-section of the rod is given by

$$\Phi = 2\pi \int_0^b Br \, dr = \pi b^2 \bar{B} \quad (4.8)$$

98  *Eddy current principles*

where $\bar{B}$ is the mean flux density in the cross-section of the rod. From equations (4.7) and (4.8) we thus have

$$\bar{B} = [2\mu_r B_0/b^2 J_0(kb)] \int_0^b r J_0(kr) \, dr \quad (4.9)$$

It can be shown (equation C.7) that

$$\int_0^b r J_0(kr) \, dr = [(r/k)J_1(kr)]_0^b = (b/k)J_1(kb) \quad (4.10)$$

from which

$$\bar{B} = (2\mu_r B_0/kb) J_1(kb)/J_0(kb) \quad (4.11)$$

where $J_1$ indicates a Bessel function of the first order.

### 4.2.2 Non-ferromagnetic rods

We now define a variable $\bar{\mu} = \bar{B}/B_0$, which Förster called the effective relative permeability. For a non-ferromagnetic rod, $\mu_r = 1$ and equation (4.11) becomes

$$\bar{\mu} = (2/kb) J_1(kb)/J_0(kb) \quad (4.12)$$

Since $kb$ is a complex quantity, $\bar{\mu}$ has both real and imaginary components $\bar{\mu}_R$ and $\bar{\mu}_I$, respectively. Furthermore, for a constant current and frequency, the rate of change of magnetic flux through a coil at any time is proportional to the induced EMF (equation 2.18) and $\bar{B}$ is then proportional to the impedance $Z = R + j\omega L$ of the coil with the rod in position, and $B_0$ to the impedance $Z_0 = j\omega L_0$ of the coil with the rod absent (section 4.1). The ratio $Z/Z_0$ is defined as the normalized impedance and is given by

$$Z/j\omega L_0 = \omega L/\omega L_0 - jR/\omega L_0 \quad (4.13a)$$

from which it can be seen that

$$\bar{\mu}_R = \omega L/\omega L_0 \quad \text{and} \quad \bar{\mu}_I = R/\omega L_0 \quad (4.13b)$$

It has been assumed that the ohmic resistance of the coil is zero, but in practice it has a small but finite value which should be subtracted from the measured resistance of the coil.

The components of the complex functions $J_0(kb)$ and $J_1(kb)$ are given by the following expressions (equations C.9)

$$J_0(kb) = J_0(j^{3/2}\beta) = \text{ber}\,\beta + j\,\text{bei}\,\beta \quad (4.14a)$$

$$J_1(kb) = -J_0'(j^{3/2}\beta) = -j^{-3/2}(\text{ber}'\beta + j\,\text{bei}'\beta) \quad (4.14b)$$

where $\beta = |kb|$.

## Coils encircling defect-free metal rods

Here we have to take into account that $kb = -j^{1/2}\beta = j^{3/2}\beta$, remembering that $-j = j^3$. Substituting equations (4.14) into equation (4.12) gives

$$\bar{\mu} = (2/\beta)(\text{bei}'\beta - j\,\text{ber}'\beta)/(\text{ber}\,\beta + j\,\text{bei}\,\beta) \tag{4.15}$$

Hence

$$\bar{\mu}_R = (2/\beta)(\text{ber}\,\beta\,\text{bei}'\beta - \text{bei}\,\beta\,\text{ber}'\beta)/(\text{ber}^2\beta + \text{bei}^2\beta) = \omega L/\omega L_0 \tag{4.16a}$$

$$\bar{\mu}_I = (2/\beta)(\text{bei}\,\beta\,\text{bei}'\beta + \text{ber}\,\beta\,\text{ber}'\beta)/(\text{ber}^2\beta + \text{bei}^2\beta) = -R/\omega L_0 \tag{4.16b}$$

Allowance should be made for the cross-section of the rod being smaller than the cross-section of the coil. This would be inevitable in practice because of the need to wind the coil on a former made from an insulating material. The fact that a multiwound coil has an effective radius greater than the inner radius should also be taken into account. The ratio of the cross-sectional area $\pi b^2$ of the rod to the effective cross-sectional area $\pi r_0^2$ of the coil is called the fill-factor $\eta$, given by

$$\eta = b^2/r_0^2 \tag{4.17}$$

where $r_0$ is the effective radius of the coil. The magnetic flux $\Phi_G$ in the air gap between the coil and the rod is given by

$$\Phi_G = \pi(r_0^2 - b^2)B_0 \tag{4.18}$$

The total flux $\Phi_T$ threading the coil is thus

$$\Phi_T = \Phi + \Phi_G = \pi r_0^2 \bar{B}_M \tag{4.19}$$

where $\bar{B}_M$ is the mean flux density corresponding to $\Phi_T$. Substituting equations (4.8) and (4.18) into equation (4.19) gives

$$\pi r_0^2 \bar{B}_M = \pi b^2 \bar{B} + \pi(r_0^2 - b^2)B_0 \tag{4.20}$$

Dividing equation (4.20) throughout by $B_0 \pi r_0^2$ we obtain

$$\bar{\mu}_M = \eta\bar{\mu} + 1 - \eta \tag{4.21}$$

where $\bar{\mu}_M$ is the effective relative permeability for any value of fill-factor. The general equations expressing the real and imaginary components of normalized impedance are thus (cf. equation 4.13b):

$$\omega L/\omega L_0 = \eta\bar{\mu}_R + 1 - \eta \tag{4.22a}$$

$$R/\omega L_0 = \eta\bar{\mu}_I \tag{4.22b}$$

The components of $\bar{\mu}$ can be derived from Bessel function tables (McLachlan, 1934) or obtained directly from the comprehensive tables

**Table 4.1** Variations of real $\bar{\mu}_R$ and imaginary $\bar{\mu}_I$ components of effective relative impedances with normalized frequency $\beta^2$

| $\beta^2(f_0)$ | $\bar{\mu}_R$ | $\bar{\mu}_I$ |
|---|---|---|
| 0 | 1.000 | 0 |
| 1 | 0.9798 | 0.1215 |
| 2 | 0.9255 | 0.2244 |
| 4 | 0.7738 | 0.3449 |
| 6 | 0.6361 | 0.3770 |
| 8 | 0.5365 | 0.3693 |
| 10 | 0.4678 | 0.3494 |
| 12 | 0.4194 | 0.3281 |
| 15 | 0.3697 | 0.3001 |
| 20 | 0.3178 | 0.2656 |
| 25 | 0.2841 | 0.2416 |
| 30 | 0.2593 | 0.2237 |
| 40 | 0.2245 | 0.1979 |
| 50 | 0.2006 | 0.1795 |
| 60 | 0.1830 | 0.1655 |
| 70 | 0.1694 | 0.1545 |
| 80 | 0.1584 | 0.1454 |
| 90 | 0.1493 | 0.1378 |
| 100 | 0.1416 | 0.1312 |
| 125 | 0.1264 | 0.1186 |
| 150 | 0.1154 | 0.1089 |
| 175 | 0.1069 | 0.1012 |
| 200 | 0.1000 | 0.0950 |
| 250 | 0.0894 | 0.0855 |
| 300 | 0.0816 | 0.0783 |
| 400 | 0.0707 | 0.0682 |
| 500 | 0.0632 | 0.0613 |
| 750 | 0.0516 | 0.0503 |
| 1 000 | 0.0447 | 0.0473 |
| 1 500 | 0.0365 | 0.0358 |
| 2 000 | 0.0316 | 0.0311 |
| 3 000 | 0.0258 | 0.0255 |
| 5 000 | 0.0200 | 0.0198 |
| 10 000 | 0.0141 | 0.0140 |
| 20 000 | 0.0100 | 0.0100 |
| 50 000 | 0.0063 | 0.0063 |
| 100 000 | 0.0045 | 0.0045 |
| 500 000 | 0.0020 | 0.0020 |
| 1 000 000 | 0.0004 | 0.0004 |
| 10 000 000 | 0.0001 | 0.0001 |

computed by Förster (1986). However, the analysis can be performed numerically using a fairly short and relatively simple program in BASIC which can be handled rapidly by most personal computers (Appendix D.1). Table 4.1 lists values of some components of $\bar{\mu}$

calculated using this program for different normalized frequencies $f_0$, where

$$f_0 = \beta^2 = \omega\mu_r\mu_0\sigma b^2 \qquad (4.23a)$$

Here $\mu_r$ is equal to unity.

An important feature of Förster's impedance analysis is the use of normalization in evaluating the components of impedance of the coil (equations 4.13 and 4.22), the radius or diameter (the fill-factor $\eta$) and the frequency (equation 4.23a) in dimensionless forms. Förster defined a 'limiting' or 'boundary' frequency $f_g$, such that $f_0 = f/f_g$, where $f$ is the operating frequency:

$$f_g = 2/\pi\mu_r\mu_0\sigma b^2 \qquad (4.23b)$$

The great advantage of normalization is the components of impedance of a coil of any radius encircling a defect-free metal cylinder of any radius can be evaluated from a single family of curves (e.g. Figure 4.2 which shows the variations of $\omega L/\omega L_0$ with $R/\omega L_0$ for non-ferromagnetic materials and for different values of $f_0$ and $\eta$, as obtained from equations 4.22). The magnitudes of $f_0$ are marked on the curves and, for given values of $\mu_r$, $\sigma$ and $b$, they are proportional to the frequencies. However, for a fixed frequency, the $f_0$ curve indicates variations of $\sigma$, when $\mu_r$ and $b$ are constant, and is often called the $\delta\sigma$ curve. Also, if the radius of the coil is constant, the fill-factor curve indicates values of the diameter $D$ of the cylinder and is called the $\delta D$ curve. Changes in both $\sigma$ and $D$ can therefore be determined simultaneously.

Figure 4.2 shows that the change in impedance with $\sigma$ decreases as the value of frequency rises, which indicates that eddy current measurements of electrical conductivity are more sensitive at lower frequencies. However, the phase difference between the $\delta\sigma$ (or $\delta f_0$) and $\delta D$ curves increases from a minimum at low frequencies to a more or less constant value of around 45° when $f_0$ exceeds 15, so the optimum frequency for conductivity measurements corresponds to $f_0 = 15$. The effects of fill-factor or diameter changes can be eliminated by phasing out that component of impedance in the direction of the $\delta D$ curve. Changes in diameter can be measured to a higher degree of sensitivity in the upper frequency range, i.e. when $f_0$ is greater than 15. These phenomena depend on the fact that the penetration depth (equation 2.90) decreases with rising frequency.

The increase of phase difference between the $\delta\sigma$ and $\delta D$ curves with frequency for lower values of $f_0$ indicates that changes in conductivity can also be determined by measuring this phase difference.

In practice both the coil and rod have finite lengths, so discrepancies arise between the measured and theoretical values of the normalized components of impedance. Provided the coil is shorter than the rod

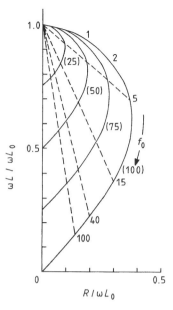

**Figure 4.2** Variation of components $\omega L/\omega L_0$ and $R/\omega L_0$ (equations 4.22) of the normalized impedance of a long coil encircling a long non-ferromagnetic (or saturated ferromagnetic) electrically conducting cylindrical rod for a normalized frequency $f_0$ and percentage fill-factor $\eta$ (in parentheses).

and the length of the coil exceeds its diameter by a factor of 2 or more, the discrepancies should not exceed a few per cent. More accurate predictions of impedance can be obtained for coils of finite length with the use of a more complicated analysis (Dodd *et al.* 1974). However, the main purpose of plotting the impedance curves is to assess optimum frequencies for testing and Förster's method is normally adequate for this purpose, even for comparatively short coils.

### 4.2.3 Ferromagnetic rods

A serious problem which may arise with the eddy current testing of ferromagnetic materials is the production of hysteresis resulting from the alternating current in the exciting coil, which gives rise to cyclical variations of the magnetic permeability $\mu_r$. This difficulty may be avoided by applying a high direct magnetic field to produce saturation (section 2.7.1) when, as explained below, the sample can be treated as non-ferromagnetic.

With non-saturated ferromagnetic metals, provided the amplitude of the current through the exciting coil is minimal, the test sample is subjected to a minor hysteresis loop at the origin of the $B-H$ curve, which reduces to a straight line when the maximum value of $H$ is negligible. The appropriate value of $\mu_r$ is the incremental permeability about the origin, known as the recoil permeability $\mu_{rec}$. When a ferromagnetic material is saturated, the origin of the minor hysteresis loop lies on the saturation part of the $B-H$ curve, which is linear and where the relative incremental permeability is equal to unity. The impedance analysis is then the same as for non-ferromagnetic metals.

Equation (4.11) shows that the normalized components of impedance for a ferromagnetic conducting rod are obtained simply by multiplying the real and imaginary components of $\bar{\mu}$ in equations (4.22) by $\mu_r$ to obtain the following generalized expressions:

$$\omega L/\omega L_0 = \eta \mu_r \bar{\mu}_R + 1 - \eta \quad (4.22\text{a}^*)$$

$$R/\omega L_0 = \eta \mu_r \bar{\mu}_I \quad (4.22\text{b}^*)$$

Equations (4.23) remain unchanged because both $f_0$ and $f_g$ are already expressed in terms of $\mu_r$.

The impedance components expressed in these equations depend on three variables, and if a two-dimensional display is required, one of them must be kept constant. Figure 4.3 shows the variations of $\omega L/\omega L_0$ and $R/\omega L_0$ for ferromagnetic cylindrical rods, with normalized frequency and relative magnetic permeability for a fill-factor $\eta$ of 90%.

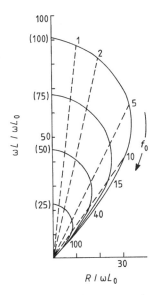

**Figure 4.3** Variation of components $\omega L/\omega L_0$ and $R/\omega L_0$ (equations 4.22*) of the normalized impedance of a long coil encircling a long unsaturated ferromagnetic conducting cylindrical rod for a normalized frequency $f_0$ and relative magnetic permeability $\mu_r$ (in parentheses) at a fill-factor $\eta = 90\%$.

## 104  Eddy current principles

The flux $\Phi$ passing through the rod is generally considerably greater than the flux $\Phi_G$ through the gap between the coil and rod, therefore the fill-factor has only a minimal influence on the impedance curves for high values of $\mu_r$. The $f_0$ curve now depends on the product $\sigma\mu_r$ and there is a decrease in the impedance change $\delta Z$ for a given value of $\delta(\sigma\mu_r)$ as the frequency increases. And the $\delta\mu_r$ curves converge for higher values of $f_0$, which suggests the changes in magnetic permeability exert less influence on values of coil impedances at higher frequencies.

### 4.3 COILS ENCIRCLING DEFECT-FREE METAL TUBES

The analysis, also originally produced by Förster (1986), for deriving the normalized impedance components of an infinitely long coil encircling an infinitely long cylindrical defect-free metal tube is considerably more complex than that for rods because boundary conditions must be applied to both the external and internal surfaces. The applicable solution to Bessel's equation for the flux density $B$ is given by equation (C.3).

$$B = A_1 J_0(j^{3/2}|kr|) + B_1 K_0(j^{1/2}|kr|) \qquad (4.24)$$

where $K_0$ is a zeroth-order Bessel function of the third kind (Appendix C) and $|k| = (\omega\sigma\mu_r\mu_0)^{1/2}$.

The treatment given below is based on work published by Bareham (1960) with respect to the values of the real and imaginary components of current densities at the inner and outer surfaces of conducting cylindrical tubes. It is adapted here to derive the values of the normalized impedance components $\omega L/\omega L_0$ and $R/\omega L_0$ of the coil encircling this kind of tube.

Let $a$ and $b$ represent, respectively, the internal and external radii of a cylindrical tube, assumed at present to be a non-ferromagnetic conductor, and $B_0$ the magnitude of the magnetic flux density at the outer surface. Equation (4.24) becomes

$$B_0 = A_1 J_0(j^{3/2}\beta) + B_1 K_0(j^{1/2}\beta) \qquad (4.25)$$

where $\beta = |kb|$. At any radial distance $r$ from the axis of the tube, the magnetic flux density $B$ is given by equation (4.24) and the electric field intensity by

$$E = -(1/\mu\sigma)\partial B/\partial r \qquad (4.26)$$

This equation arises from curl $\mathbf{H} = \mathbf{J} = \sigma\mathbf{E}$ (i.e. equation (2.24) and Ohm's law) as applied to electrical conductors and expressed here in cylindrical polar coordinates and assuming axisymmetric conditions.

Both $B$ and $E$ are functions only of $r$ but $E$ acts in the $r, \phi$ plane. From (4.24) and (4.26), we have

$$E = -(|k|j^{1/2}/\sigma\mu)[A_1 j J_0'(j^{3/2}|kr|) + B_1 K_0'(j^{1/2}|kr|)] \quad (4.27)$$

The expression curl $E = -\partial B/\partial t$ (equation 2.21) can now be written as

$$(1/r)\partial(rE)/\partial r = -\partial B/\partial t = -j\omega B \quad (4.28)$$

since $B$ varies sinusoidally with time. Integrating equation (4.27) and dropping the integration constant, since $E$ must be finite for *all* values of $r$, we have

$$E = -j\omega r B/2 \quad (4.29)$$

At the inner surface, $r = a$ and $B = B_i$, and eliminating $E$ from equations (4.27) and (4.29) gives

$$(j^{1/2}/2\sigma\mu)B_i = A_1 j J_0'(j^{3/2}\alpha) + B_1 K_0'(j^{1/2}\alpha) \quad (4.30)$$

where $\alpha = |ka|$. Applying equation (4.24) produces

$$B_i = A_1 J_0(j^{3/2}\alpha) + B_1 K_0(j^{1/2}\alpha) \quad (4.31)$$

The unknown quantity $B_i$ is then eliminated from equations (4.30) and (4.31), and the resultant expression is solved simultaneously with equation (4.25) to give the following values of $A_1$ and $B_1$:

$$A_1 = B_0 G/D \quad \text{and} \quad B_1 = B_0 F/D \quad (4.32)$$

where

$$D = G J_0(j^{3/2}\beta) + F K_0(j^{1/2}\beta)$$
$$F = j J_0'(j^{3/2}\alpha) - (j^{1/2}\alpha/2) J_0(j^{3/2}\alpha)$$
$$G = (j^{1/2}\alpha/2) K_0(j^{1/2}\alpha) - K_0'(j^{1/2}\alpha)$$

The complex functions $J_0$, $K_0$, $J_0'$ and $K_0'$ are each then resolved into their real and imaginary components, as shown in equations (C.9), (C.18) and (C.19), and the quantities $G/D$ and $F/D$ rationalized to give

$$(G/D) = (G/D)_R + j(G/D)_I \text{ and } (F/D) = (F/D)_R + j(F/D)_I \quad (4.33)$$

where the subscripts R and I indicate the respective real and imaginary components.

From equation (4.24) the magnetic flux $\Phi$ passing through any cross-section of the tube is given by

$$\Phi = 2\pi \int_a^b Br\,dr = 2\pi B_0\{(G/D)[(a^2/j^{3/2}\alpha)J_0'(j^{3/2}\alpha) - (b^2/j^{3/2}\beta)J_0'(j^{3/2}\beta)]$$
$$+ (F/D)[(a^2/j^{1/2}\alpha)K_0'(j^{1/2}\alpha) - (b^2 j^{1/2}\beta)K_0'(j^{1/2}\alpha)]\} = \pi(b^2 - a^2)\bar{B}$$
$$(4.34a)$$

where $\bar{B}$ represents the effective value of $B$ (cf. equation 4.8). The complex functions in equation (4.34a) are each resolved, as before, and rationalized to produce the expression

$$P = P_R + jP_I$$

Equation (4.34a) then takes the form

$$\Phi = 2\pi B_0(P_R + jP_I) = \pi(b^2 - a^2)\bar{B} \tag{4.34b}$$

from which

$$\bar{B}/B_0 = \bar{\mu} = 2(P_R + jP_I)/(b^2 - a^2) \tag{4.34c}$$

For a ferromagnetic material, $\bar{B}$ and therefore the effective relative permeability $\bar{\mu}$ are multiplied by the relative permeability $\mu_r$ (section 4.2.3).

Account must now be taken of the space enclosed by the inner surface of the tube, where the flux density has a constant and complex value $B_i$, expressed by equation (4.31), for which the value of the flux $\Phi_i$ is given by

$$\Phi_i = \pi a^2 B_i = \pi a^2 \mu_i B_0 \tag{4.35a}$$

where $\mu_i$ is the corresponding relative permeability

$$\mu_i = B_i/B_0 = \mu_{iR} + j\mu_{iI} \tag{4.35b}$$

For a fill-factor of less than 100%, the flux $\Phi_G$ between the coil and the outer surface is equal to $\pi(r_0^2 - a^2)B_0$, where $r_0$ is the radius of the coil. Equating the sum of the fluxes $\Phi$, $\Phi_i$ and $\Phi_G$ to the total flux $\Phi_T$ threading the coil gives

$$\pi r_0^2 \bar{B}_M = \pi(b^2 - a^2)\mu_r \bar{B} + \pi a^2 B_i + \pi(r_0^2 - b^2)B_0 \tag{4.35c}$$

where $B_M$ is the effective flux density through the cross-section of the coil (cf. equation 4.20). Dividing both sides of this equation by $\pi r_0^2$, putting $\bar{\mu}_M = \bar{B}_M/B_0$ and the fill-factor $\eta = (b/r_0)^2$ (cf. equation 4.21), the total (effective relative permeability) is then given by

$$\bar{\mu}_M = 2\eta\mu_r\bar{\mu}/b^2 + \eta p^2 \mu_i + 1 - \eta \tag{4.36a}$$

where $p = a/b$, i.e. the radius or diameter ratio. The real and imaginary components of the normalized impedance of the coil are thus

$$\omega L/\omega L_0 = \eta(\mu_r\bar{\mu}_R/b^2 + p^2\mu_{iR}) + 1 - \eta \tag{4.36b}$$

$$R/\omega L_0 = \eta(\mu_r\bar{\mu}_r/b^2 + p^2\mu_{iI}) \tag{4.36c}$$

But beware, the calculation of these quantities can be a formidable task. However, it is very much simplified with the use of a personal computer. A program in BASIC written for this purpose is given in Appendix D.2.

# Coils encircling defect-free metal tubes

Setting $\beta^2 = \omega\mu_r\mu_0\sigma b^2$ as the normalized frequency $f_0$, as before, impedance curves can be plotted in a manner similar to that for cylindrical rods, but because the components of $\mu$ depend on the ratio $p = a/b$, a separate set of curves is required for each value of $p$. To simplify the program, it is necessary for $b$ to appear explicitly in equations (4.36), but being a multiplying factor forming part of the expression $P$ (equations 4.34), it is automatically eliminated in the final expressions and identical results are thus obtained for a given ratio $a/b$, irrespective of the value of $b$.

Figure 4.4 shows the variations of normalized impedance components with normalized frequency and fill-factor for a non-ferromagnetic cylindrical tube with a diameter ratio of 0.75. Comparing these curves with those for a non-ferromagnetic cylindrical rod (Figure 4.2), it can be seen that at very high frequencies, where through-penetration to the inner surface is negligible, the two sets of curves coincide. Figure 4.5 compares the curves showing the variations of $f_0$ for diameter ratios of 0.1 and 0.9. As would be expected, the curves coincide for high values of normalized frequency.

As before, allowance must be made for the fact that both coil and tube are finite in length; allowance must also be made for the increase in fill-factor caused by the finite thicknesses of the coil former and windings.

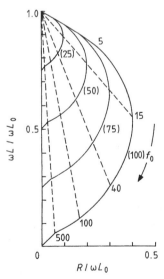

**Figure 4.4** Variation of components $\omega L/\omega L_0$ and $R/\omega L_0$ (equations 4.36) of the normalized impedance of a long coil encircling a long non-ferromagnetic (or saturated ferromagnetic) electrically conducting cylindrical tube for a normalized frequency $f_0$ and percentage fill-factor $\eta$ (in parentheses). The diameter ratio is 75%.

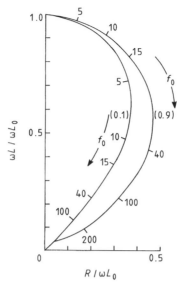

**Figure 4.5** Variation of components $\omega L/\omega L_0$ and $R/\omega L_0$ (equations 4.36) of the normalized impedance of a long coil encircling a long non-ferromagnetic electrically conducting cylindrical tube for normalized frequencies $f_0$ at diameter ratios of 0.1 and 0.9. The fill-factor is 100% and the $f_0$ values are shown on each curve.

## 4.4 INTERNAL COAXIAL COILS IN DEFECT-FREE METAL TUBES

The analysis for an internal coaxial coil inside a defect-free metal tube can be more easily followed with reference to Figure 4.6, which shows the continuity of the lines of flux. The lines diverge on leaving the tube and they curve round so as to pass mostly through the material of the tube and any intervening space between the outer windings of the coil and the inner surface of the tube. They then converge to re-enter the coil at the opposite end. For a long coil and tube it can be assumed the lines of flux are parallel for most of the region under consideration and the impedance analysis can be performed with reference to section 4.3.

The effect of inducing eddy currents in the tube is to reduce the value of the magnetic flux $\Phi'$ through the coil for a given current and frequency, in the absence of the sample, so that it becomes $\Phi_M$ with the coil in position, for the same current and frequency. If $\phi$ is the flux inside the metal and $\Phi_i$ the flux in the space between the outer windings of the coil and the inner surface of the tube, we have

$$\Phi_M = \Phi' - \Phi - \Phi_i \tag{4.37a}$$

# Coils scanning the surfaces of defect-free conductors

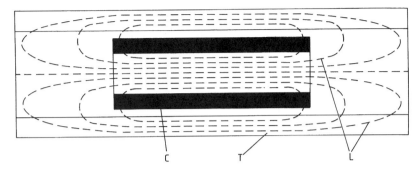

**Figure 4.6** Configuration of magnetic flux lines (L) induced by an internal coil (C) coaxial with an electrically conducting cylindrical tube (T).

Here $\Phi_M = \pi b^2 B_M$, $\Phi' = \pi r_0^2 B_0$, $\Phi = \pi(b^2 - a^2)\mu_r \bar{B}$ and $\Phi_i = \pi(a^2 - r^2)B_i$, where the symbols are defined in section 4.3. By dividing equation (4.37a) by $2\pi r_0 B_0$, it can be seen that

$$B_M/B_0 = 1 - [(b^2 - a^2)/r_0^2](\mu_r \bar{B}/B_0) - [(a^2 - r_0^2)/r_0^2](B_i/B_0) \quad (4.37b)$$

Substituting the effective relative permeabilities gives

$$\mu_r = 1 - \eta'[(1/p^2) - 1]\bar{\mu} - (\eta' - 1)\mu_i \quad (4.38)$$

where $p = a^2/b^2$, as before, and $\eta' = a^2/r_0^2$, i.e. the fill-factor. Resolving the right-hand side of equation (4.38) into its real and imaginary parts and substituting the normalized impedance of the coil for $\mu_r$ yields

$$\omega L/\omega L_0 = 1 - \eta'[(1/p^2) - 1]\bar{\mu}_R - (\eta' - 1)\mu_{iR} \quad (4.39a)$$

$$R/\omega L_0 = \eta'[(1/p^2) - 1]\bar{\mu}_I - (\eta' - 1)\mu_{iI} \quad (4.39b)$$

The computer program (Appendix D.2) applicable to coils encircling defect-free tubes can easily be modified to evaluate equations (4.39). As before, the analysis decreases in reliability with decrease in coil length.

## 4.5 COILS SCANNING THE SURFACES OF DEFECT-FREE CONDUCTORS

### 4.5.1 General considerations

The theoretical treatment given here applies to an air-cored coil which scans the surface of a conducting material with its axis perpendicular to it. The coil has a finite length and its windings are of negligible thickness, which can be achieved in practice with only one or two layers of fine-gauge wire, so the value of the effective radius $r_0$ is very close to the internal radius. A typical coil of this kind has a radius of 1 mm, a length of between 0.5 and 2 mm and an inductance of 10–40 µH, with an optimum range of excited frequencies extending from 200 kHz to over 2 MHz.

In practice the coils are usually wound on formers or rods to provide stability. Provided the formers or rods are made from dielectrics having magnetic permeabilities virtually equal to the permeability of air, the coils can be regarded effectively as being air-cored. The inductance of a coil at a given frequency can be considerably increased by introducing a ferrite core, which has a high magnetic permeability and a low electrical conductivity (e.g. of the order of 100 S m$^{-1}$), thus increasing the sensitivity of measurement. This is advantageous when testing non-ferromagnetic conductors. The author has observed that when a ferrite core is placed inside an air-cored coil there is no measurable change in its normalized impedance at a given frequency and for a given lift-off above a non-ferromagnetic conductor, provided the current through the coil is low enough to avoid the small but measurable hysteresis effects in the ferrite. This may not be the case when testing ferromagnetic conductors because of an increase in magnetic permeability arising from the greater amount of magnetic flux induced by the coil. The increased magnetic field resulting from this flux is normally sufficient to produce a measurable hysteresis in the ferrite, hence non-linearities and the consequent introduction of harmonic frequencies in the potential differences across the coil.

The analysis given below can be applied, with some decrease in accuracy, e.g. a few per cent, to short flat coils with multilayer windings, i.e. pancake coils, if it is assumed that the effective radius is the mean of the internal and external radii (Dodd and Deeds, 1968, Smith and Dodd, 1975). This section concerns only surface-scanning coils excited at frequencies high enough to avoid any measurable penetration of eddy currents to the lower surface of the sample. The effective penetration ranges from about 3–5 times the skin depth $\delta$ (equation 2.90) depending on the quality of equipment used and the signal-to-noise ratio of the detector, which can be varied by using its sensitivity control. The reader who is interested in the effects of through-penetration should note the final paragraph in section 4.5.2. The analysis given here for the surface-scanning air-cored probe coil differs considerably from the analyses relevant to encircling and internal axial coils. This mainly arises from the poor flux linkage between the coil and the material being tested. Fortunately, however, axial symmetry can still be assumed. The parameters affecting the values of the normalized components of impedance of the coil are the normalized frequency $f_0$ and normalized lift-off, as defined below. Since there is no measurable dimension which is characteristic of a sample in which there is no through-penetration, the dimension component of $f_0$ is taken as the effective radius $r_0$ of the coil (see above). Thus $f_0 = \beta^2 = \omega\mu\sigma r_0^2$, where the symbols have been defined previously (section 4.2.1).

The lift-off $h$ of the coil is the height of its base above the surface of the sample and its normalized value $k$ is equal to the ratio $h/r_0$. In determining the value of $h$, account must be taken of the fact that the actual length of the coil is greater than the effective length by a factor of about 20%, as a result of the divergence of the lines of magnetic flux at its ends. And the former of a coil must extend slightly beyond the ends of the windings, usually by about 0.2 mm, to keep them in position. These factors are important because of the comparatively high changes in impedance of coils with increase in $h$ for small changes of lift-off. Separate sets of normalized impedance curves must be plotted for coils having different values of the normalized coil length $k_0$, where $k_0 = l_0/r_0$.

### 4.5.2 Impedance analysis

This section obtains the normalized components of impedance of a coil scanning the surface of a defect-free electrical conducting material, either ferromagnetic or non-ferromagnetic. Dodd and Deeds (1968) derived values for these components, but only for a single-turn coil and a non-ferromagnetic conductor. They initially considered the variations of the magnetic vector potential $A$ in an electromagnetic field, rather than the flux density $B$, with the current density $J$ in a single-turn delta-function, i.e. filamentary, coil. The advantage of using $A$ (equation 2.25) is that its axial symmetry is the same as for $J$; the z-axis coincides with the axis of the coil. All functions of the coordinate $\phi$ remain constant. Consequently, only the cylindrical coordinates $(r, z)$ need be used and they are applicable to all parts of the filamentary coil of radius $r_0$. Dodd and Deeds assumed a linear, isotropic and homogeneous medium in which the magnetic permeability $\mu$ is constant and for which

$$\partial^2 A/\partial r^2 + (1/r)\partial A/\partial r + \partial^2 A/\partial z^2 - A/r^2 - j\omega\mu\sigma A = -\mu J \quad (4.40)$$

This equation is expressed in cylindrical coordinates. $A$ and $J$ are the $\theta$ components of $A$ and $J$ respectively, and are functions of $r$ and $z$ only; $J$ is assumed to vary sinusoidally (i.e. as $\exp j\omega t$). $\sigma$ is the electrical conductivity and $\omega$ the angular frequency. This equation is applicable to frequencies lower than 10 MHz where any function of the electrical permittivity can be neglected. Dodd and Deeds assumed the filamentary coil had a rectangular cross-section with dimension $\delta(r - r_0)$ and $\delta(z - z_0)$. For a total current $I$, equation (4.40) becomes

$$\partial^2 A/\partial r^2 + (1/r)\partial A/\partial r + \partial^2 A/\partial z^2 - A/r^2 - j\omega\mu\sigma A$$
$$= -\mu I/\delta(r - r_0)\delta(z - z_0) \quad (4.41)$$

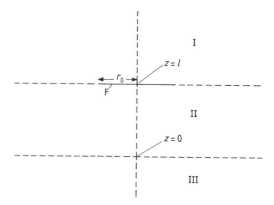

**Figure 4.7** Horizontal filamentary (delta-function) coil (F) located in air above the horizontal surface of an electrical conductor. Regions I, II and III are also indicated (see text).

A typographical error by Dodd and Deeds in the term on the right-hand side of equation (4.41) is corrected here. This error does not affect the final result because it cancels when the impedance of the coil is normalized.

Dodd and Deeds considered the single-turn filamentary coil to be located in air above two parallel-sided non-ferromagnetic conductors having plane horizontal surfaces. Here we need consider only a single conductor, either ferromagnetic or non-ferromagnetic in a manner similar to Hajian *et al.* (1983). The field of interest (Figure 4.7) is divided up into three regions, i.e. (I) in air above the delta-function coil at $z = 1$, (II) in air between the coil and the surface of the conductor at $z = 0$ and (III) in the conductor. The equations of propagation in terms of $A$ are as follows.

Regions I and II

$$\partial^2 A/\partial r^2 + (1/r)\partial A/\partial r + \partial^2 A/\partial z^2 - A/r^2 = 0 \qquad (4.42)$$

where the value of $\sigma$ is negligible for air, and

Region III

$$\partial^2 A/\partial r^2 + (1/r)\partial A/\partial r + \partial^2 A/\partial z^2 - A/r^2 - j\omega\mu_1\sigma_1 A = 0 \qquad (4.43)$$

where $\mu_1$ and $\sigma_1$ are the respective values of $\mu$ and $\sigma$ for the conductor. Equation (4.43) is solved by separating the functions of $r$ and $z$, i.e. putting $A = A(r, z) = A(r)A(z)$; dividing by this product gives

$$[1/A(r)]\partial^2 A(r)/\partial r^2 + [1/rA(r)]\partial A(r)/\partial r$$
$$+ [1/A(z)]\partial^2 A(z)/\partial z^2 - 1/r^2 - j\omega\mu_1\sigma_1 = 0 \quad (4.44)$$

Putting $[1/A(z)]\partial^2 A(z)/\partial z^2$ equal to a separation 'constant' $\alpha_1^2$ defined as

$$\alpha_1^2 = \alpha^2 + j\omega\mu_1\sigma_1 \qquad (4.45)$$

we have

$$A(z) = A' \exp(\alpha_1 z) + B \exp(\alpha_2 z) \qquad (4.46)$$

where $A'$ and $B$ are 'constants'. Equation (4.44) then becomes

$$[1/A(r)]\partial^2 A(r)/\partial r^2 + [1/rA(r)]\partial A(r)/\partial r + \alpha^2 - 1/r^2 = 0 \qquad (4.47)$$

This is a Bessel equation of the *first* order (equation C.4), for which the general solution is

$$A(r) = CJ_1(\alpha r) + DY_1(\alpha r) \qquad (4.48)$$

where $C$ and $D$ are 'constants' and $J_1$ and $Y_1$ are respectively first-order Bessel functions of the first and second kind (equation C.5). Combining the solutions (equations 4.46 and 4.48), we have

$$A(r, z) = [A' \exp(\alpha_1 z) + B \exp(-\alpha_1 z)][CJ_1(\alpha r) + DY_1(\alpha r)] \qquad (4.49)$$

The terms $A'$, $B$, $C$ and $D$ are constant only for a given value of $\alpha$, itself a continuous variable. The complete solution is obtained by integrating this expression for values of $\alpha$ from zero to infinity

$$A(r, z) = \int_0^\infty [A'(\alpha) \exp(\alpha_1 z) + B(\alpha) \exp(-\alpha_1 z)][C(\alpha)J_1(\alpha r) + D(\alpha)Y_1(\alpha r)]d\alpha \qquad (4.50)$$

Since $Y_1(\alpha r)$ is infinite when $\alpha r$ is zero (McLachlan, 1934), the constant $D(\alpha)$ is equal to zero in all regions. The value of $A'(\alpha)$ is zero in region I where $z$ goes to $+\infty$, and $B(\alpha)$ is zero in region III where $z$ goes to $-\infty$. The values of $A(r, z)$ for each region then become

$$A^{\mathrm{I}}(r, z) = \int_0^\infty B_1(\alpha) \exp(-\alpha z) J_1(\alpha r) d\alpha \qquad (4.51a)$$

$$A^{\mathrm{II}}(r, z) = \int_0^\infty [C_2(\alpha) \exp(\alpha z) + B_2(\alpha) \exp(-\alpha z)] J_1(\alpha r) d\alpha \qquad (4.51b)$$

$$A^{\mathrm{III}}(r, z) = \int_0^\infty C_3(\alpha) \exp(\alpha_1 z) J_1(\alpha r) d\alpha \qquad (4.51c)$$

We now depart from the treatment of Dodd and Deeds. Originally Hajian and Blitz (1986) continued the treatment to derive expressions for the components of the normalized impedances of a coil located above a non-ferromagnetic conductor. However, further work (Oaten, 1989) considered the general case of both ferromagnetic and non-ferromagnetic conductors. Account was taken of the fact that when changes occur in $\mu$, as between regions II and III, the boundary conditions depend on continuity of the magnetic field intensity $H$, not the vector potential $A$; this is due to the changes in the relative magnetic permeability $\mu_r$. Thus $B/\mu_r$ is continuous, hence $A/\mu_r$ is continuous. At $z = l$ we have

$$A^I(r, l) = A^{II}(r, l) \tag{4.52a}$$

and

$$(\partial/\partial z)A^I(r, z)_{z=l} = (\partial/\partial z)A^{II}(r, z)|_{z=l} - \mu_0 I/\delta(r - r_0) \tag{4.52b}$$

which shows how the current through the coil contributes to the continuity. At $z = 0$ we have

$$A^{II}(r, 0) = A^{III}(r, 0) \tag{4.52c}$$

and

$$(\partial/\partial z)A^{II}(r, z)|_{z=0} = (1/\mu_r)(\partial/\partial z)A^{III}(r, z)|_{z=0} \tag{4.52d}$$

The value of $A$ at the delta-function coil is obtained by calculating $A^I$ and $A^{II}$ then combining them. It is first necessary to evaluate the 'constants' $B_1(\alpha)$, $B_2(\alpha)$, $C_2(\alpha)$ and $C_3(\alpha)$ from these boundary conditions. This can be achieved in a relatively simple manner, as used by Dodd and Deeds: substitute the relevant equations (4.51) into equations (4.52), multiply both sides of equations (4.52) by an operator $\int_0^\infty J_1(\alpha' r) r \, dr$, reverse the order of integration then apply the Fourier–Bessel equation (e.g. Bell, 1968). Their values are

$$B_1(\alpha) = [\mu_0 I J_1(\alpha r)/2r_0]\{\exp(\alpha l) + \exp(-\alpha l)[(\mu_r \alpha - \alpha_1)/(\mu_r \alpha + \alpha_1)]\} \tag{4.53a}$$

$$B_2(\alpha) = [\mu_0 I J_1(\alpha r)/2r_0]\exp(-\alpha l)[(\mu_r \alpha - \alpha_1)/(\mu_r \alpha + \alpha_1)] \tag{4.53b}$$

$$C_2(\alpha) = [\mu_0 I J_1(\alpha r)/2r_0]\exp(-\alpha l) \tag{4.53c}$$

The derived expressions are

$$A^I(r, z) = (\mu_0 I/2r_0) \int_0^\infty J_1^2(\alpha r_0)(\exp[-\alpha(l+z)])\{\exp(2\alpha l) + [(\mu_r \alpha - \alpha_1)/(\mu_r \alpha + \alpha_1)]\}d\alpha \tag{4.54a}$$

$$A^{II}(r, z) = (\mu_0 I/2r_0) \int_0^\infty J_1^2(\alpha r_0)\{\exp[-\alpha(l+z)]\}\{\exp(2\alpha z) + [(\mu_r \alpha - \alpha_1)/(\mu_r \alpha + \alpha_1)]\}d\alpha \tag{4.54b}$$

In the above equations $J_1(\alpha r_0)$ has been substituted for $J_1(\alpha r)$ because the magnetic flux density $B$ is constant, hence $A$ is constant at all points

## Coils scanning the surfaces of defect-free conductors

inside the coil, i.e. for all values of $r$ from 0 to $r_0$. The value of $A$ at the delta-function coil is then given by

$$A = (\mu_0 I / 2r_0) \int_0^\infty J_1^2(\alpha r_0) \{\exp[\pm\alpha(l-z)]$$
$$+ [(\mu_r \alpha - \alpha_1)/(\mu_r \alpha + \alpha_1)] \exp[-\alpha(l+z)]\} d\alpha \quad (4.55)$$

Figure 4.8 illustrates the arrangement of the complete coil of length $l_0$ above the surface of the sample and a position of the filamentary coil. When $z > l$ the modulus $|l - z|$ takes a positive sign and when $z < l$ it takes a negative sign. Integrations must be performed on $A$ firstly as a function of $z$, with respect to the filament, then as a function of $l$ to provide a value for the complete coil. Putting $l_0 = m - h$ and $n_0$ as the number of turns per unit length, we define $i_0 = I/n_0 l_0$. For a length d$l$ of the coil, the corresponding current d$I$ is given by d$I = i_0 n_0$ d$l$ and the number of turns within a length d$z$ of the coil is equal to $n_0$ d$z$.

The potential difference is related to the vector potential in that the electric field intensity is the potential gradient (section 2.4.2). The potential difference $V'$ across a single turn of a coil is equal to $-\oint E \, dl$. From equation (2.26) $E = -\partial A/\partial t$ and we then have

$$V' = 2\pi r_0 \partial A/\partial t = 2j\omega\pi r_0 A \quad (4.56)$$

Thus for an incremental coil of length d$l$, the corresponding potential difference, disregarding the prime, is given by

$$dV = j\omega\pi\mu_0 n_0^2 i_0 \times (\text{integral}) \, dz \, dl \quad (4.57)$$

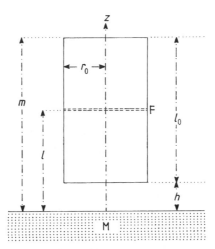

**Figure 4.8** Location of an air-cored solenoid coil (C) above the plane surface of a non-ferromagnetic electrical conductor (M). F is the delta-function coil and $z$ indicates the $z$-axis. (Reprinted, by permission, from Hajian *et al.*, 1983; copyright Butterworth-Heinemann Ltd)

where the integral referred to is that on the right-hand side of equation (4.55)

The total potential difference across the coil of length $l_0$ is

$$V = \int_{l=h}^{m}\int_{z=h}^{m} dV = j\omega\mu_0 n_0^2 i_0 \int_{l=h}^{m}\int_{z=h}^{m}\int_0^{\infty} J_1^2(\alpha r_0)\{(2l_0/\alpha)$$
$$- (2/\alpha)^2[1 - \exp(-\alpha l_0)]$$
$$+ [(\mu_r\alpha - \alpha_1)/\alpha^2(\mu_r\alpha + \alpha_1)]\exp(-2\alpha h)[1 - \exp(-\alpha l_0)]^2\}\,d\alpha\,dz\,dl$$
(4.58)

For convenience, we put $r_0\alpha = X$ and $r_0\alpha_1 = X_1$ so that

$$X_1^2 - X^2 = j\beta^2 \tag{4.59}$$

where $\beta^2$ is the normalized frequency. Normalizing the length of the coil and lift-off, respectively, so that $k_0 = l_0/r_0$ and $k = h/r_0$, and substituting $I = i_0 n_0 l_0$, we have

$$V = j\omega KI(2k_0 I_1 - 2I_2 + 2I_3 + I_4) \tag{4.60}$$

where $K = (\pi\mu_0 n_0/l_0)$ and

$$I_1 = \int_0^{\infty} [J_1^2(X)/X]\,dX \tag{4.61a}$$

$$I_2 = \int_0^{\infty} [J_1(X)/X]^2\,dX \tag{4.61b}$$

$$I_3 = \int_0^{\infty} [J_1(X)/X]^2[\exp(-k_0)]\,dX \tag{4.61c}$$

$$I_4 = \int_0^{\infty} [J_1(X)/X]^2[\exp(-2kX)][1 - \exp(-k_0 X)]^2 f(X)\,dX \tag{4.61d}$$

where $f(X) = (\mu_r X - X_1)/(\mu_r X + X_1)$. From equation (4.60) the impedance $Z$ of the coil is given by

$$Z = V/I = j\omega K(2k_0 I_1 - 2I_2 + 2I_3 + I_4) \tag{4.62}$$

For infinite lift-off, $k$ is infinite and $I_4$ vanishes, so

$$Z_0 = j\omega L_0 = j\omega K(2k_0 I_1 - 2I_2 + 2I_3) \tag{4.63}$$

Dividing this equation into equation (4.62) gives

$$Z/Z_0 = (R + j\omega L)/j\omega L_0$$

i.e.

$$\omega L/\omega L_0 - jR/\omega L_0 = 1 + I_4/(2k_0 I_1 - 2I_2 + 2I_3) \tag{4.64}$$

where $I_1$ and $I_2$ are standard integrals, given by

$$I_1 = 1/2 \quad \text{and} \quad I_2 = 4/3\pi$$

The evaluations of $I_3$ and $I_4$, hence of $\omega L/\omega L_0$ and $R/\omega L_0$, have to be performed numerically using a computer. Before this can be done, it is necessary to resolve the terms on the right-hand side of equation (4.64) into their real and imaginary components. However, because $\alpha$ and $X$ were postulated as real quantities, the only complex expression is $X_1$ (equation 4.59), which forms part of the parameter $f(X)$ contained by the integral $I_4$ in equation (4.61d). Putting $\beta^2/X^2 = \tan\theta$, it can easily be shown that

$$X_1 = Y(\cos\theta/2 + j\sin\theta/2) \qquad (4.65)$$

where $Y = [X^4 + \beta^4]^{1/2}$, from which we obtain

$$f(X) = [(\mu_r X)^2 - Y^2 - 2j\mu_r XY \sin\theta/2]/[(\mu_r X)^2 + Y^2 + 2j\mu_r XY \cos\theta/2]$$

Here

$$\begin{aligned}\sin\theta/2 &= \{[1 - (\mu_r X)^2/Y^2]/2\}^{1/2} \\ \cos\theta/2 &= \{[1 + X^2/Y^2]/2\}^{1/2}\end{aligned} \qquad (4.66)$$

The normalized components of impedance can be evaluated on a personal computer using the reasonably short program given in Appendix D.3. By limiting the expansion of $J_1(X)$ to 15 terms it was possible to calculate the normalized components of impedance to four decimal places for a given normalized frequency $f_0$ in a second or so.

Since this method involves integration over a finite length of the coil, it does not suffer from the uncertainties introduced in the analyses for encircling coils (sections 4.2 and 4.3), where the lengths of coil and sample are assumed to be infinite. By correcting for coil length and lift-off, close agreement to three decimal places between predicted and measured values and almost perfect agreement to two decimal places have been obtained for non-ferromagnetic materials at values of $f_0$ between 200 and 800 at all values of lift-off for a coil having a normalized length $k_0$ of 0.6 (Blitz, 1989).

However, for values of $k_0$ greater than 1.5 and $k$ greater than 0.7, with non-ferromagnetic conductors, a good approximation can be obtained when the normalized frequency $f_0$ is between the approximate limits of 600 and 3000, which usually covers frequencies within the upper kilohertz range, using a 5 mm radius air-cored coil. In these circumstances it has been shown (Hajian *et al.*, 1983) that

$$I_3 \approx 1/4k_0$$

and

$$(X - X_1)/(X + X_1) \approx -1 + \sqrt{2}(1-j)X/\beta$$

so that

$$I_4 \approx (k_0^2/8P)[Q(1-j)/\sqrt{2}\beta P - 1]$$

giving

$$\omega L/\omega L_0 \approx 1 + (k_0^2/8PS)(Q/\sqrt{2}\beta P - 1) \quad (4.67a)$$

$$R/\omega L_0 \approx k_0^2 Q/(8\sqrt{2}\beta P^2 - 1) \quad (4.67b)$$

where

$$P = k(k+k_0)(2k+k_0)$$
$$Q = k_0^2 + 6k^2 + 6kk_0$$
$$S = k_0 - 8/3\pi + 1/2k$$
$$\beta = (\omega\mu_0 \sigma r_0^2)^{1/2} = f_0^{1/2}$$

Sets of impedance curves, applicable to non-ferromagnetic metals and derived from the more exact theory, for normalized coil lengths $k_0$, respectively equal to 0.5 and 1, are shown in Figures 4.9 and 4.10. They bear an apparent similarity to those for encircling coils (Figure 4.2) but with normalized lift-off $k_0$ replacing the fill-factor. Because of the much smaller degree of flux linkage between the coil and the sample than with encircling coils, the $f_0$ curves for zero lift-off do not extend to

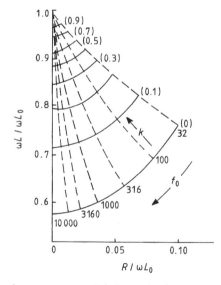

**Figure 4.9** Variation of components $\omega L/\omega L_0$ and $R/\omega L_0$ (equation 4.64) of the normalized impedance of a surface-scanning solenoid coil with its axis perpendicular to the plane surface of a non-ferromagnetic electrical conductor. The normalized frequency is $f_0$ and the normalized lift-off is $k$. The normalized length $k$ of the coil is 0.5. The $R/\omega L_0$ scale has been expanded for clarity.

# Coils scanning the surfaces of defect-free conductors

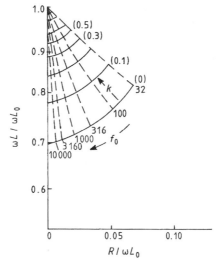

**Figure 4.10** As Figure 4.9 but this time the coil has normalized length $k_0 = 1$.

the origin. The ratio is reduced between the resistive component and the inductive component of the impedance, and for greater clarity the $R/\omega L_0$ scales have been enlarged in these figures. Furthermore, for the reasons given earlier, zero lift-off is impossible to achieve in practice. In common with the use of encircling coils, the incremental changes of the normalized components of impedance decrease as $f_0$ increases, so that the sensitivity in measuring electrical conductivity and magnetic permeability is higher at lower frequencies. There is also an increase in the phase angle between the lift-off curve and the $f_0$ curve (i.e. the $\delta\sigma$ curve) with increases in $f_0$ (i.e. $\sigma$ at a given frequency). Electrical conductivities and magnetic permeabilities can therefore be determined at a given frequency by observing the directions of the lift-off impedance curves, as described below. In practice it is not possible to display graphically the variations in the normalized value $\omega L/\omega L_0$ and $R/\omega L_0$ of the components of impedance of a surface-scanning coil, for different values of relative magnetic permeability at a given lift-off, in the same way as with an encircling coil (Figure 4.3). The curves for the different magnetic permeabilities become superimposed on one another but the values of $f_0$ for each of the curves do not coincide in position. With increases in $\mu$, the points representing the values of $f_0$ are displaced along the curve towards the origin. This is in contrast to what is observed in Figure 4.3, where the superposition takes place only at higher values of $f_0$, but this is to be expected because of the much lower flux linkage between a conductor and a surface-scanning coil than between a conductor and an encircling coil.

Figure 4.11 illustrates lift-off curves with a coil having a given size at a fixed absolute frequency, as opposed to normalized frequency, for both ferromagnetic and non-ferromagnetic metals having different values of conductivity and permeability, as obtained with the analysis given earlier and using the computer program in Appendix D.3. The curves indicate the changes in impedance components for values of lift-off ranging from zero to infinity. Note the differences between the directions of the curves for ferromagnetic and non-ferromagnetic conductors, respectively.

The analysis can be extended to apply to a coil located above two conductors. This can be left as an exercise for the interested reader, who by following the work of Dodd and Deeds (1968), should not find any undue difficulty. By replacing the lower of the two conductors with air, the analysis can be applied to a coil situated above a metal sheet which allows through-penetration. The same paper by Dodd and Deeds also contains an impedance analysis for an infinite coil encircling a two-conductor rod; this provides an alternative approach

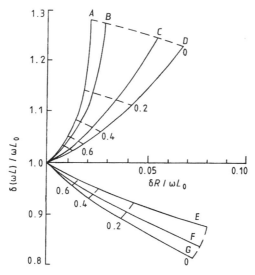

**Figure 4.11** Theoretical predictions of variations and $\delta(\omega L)/\omega L_0$ and $\delta R/\omega L_0$ (equation 4.64) from balance position at 'infinite' lift-off plotted versus height of lift-off in millimetres for a surface-scanning coil. The coil, radius 1 mm and length 1 mm, is excited at a frequency of 63.3 kHz for defect-free metal samples with values of electrical conductivity $\sigma$ and relative magnetic permeability $\mu_r$ as follows.

|  | A | B | C | D | E | F | G |
|---|---|---|---|---|---|---|---|
| $\sigma$ (MS m$^{-1}$) | 1 | 1 | 10 | 10 | 20 | 40 | 60 |
| $\mu_r$ | 100 | 50 | 100 | 50 | 1 | 1 | 1 |

# Defect modelling

to the analyses for cylindrical rods and tubes given in sections 4.2 and 4.3, but in the opinion of the author, the method of Dodd and Deeds is more difficult to follow and is no more advantageous.

## 4.6 DEFECT MODELLING

### 4.6.1 General considerations

When a defect or any other kind of discontinuity occurs in a metal object in which eddy currents are induced, the paths of the current, hence the lines of flux, are diverted in a manner characteristic of the nature of the discontinuity and changes in impedance of the coil take place. It is clear that defects, e.g. planar cracks and laminations, oriented in the same direction as the eddy currents cannot be detected and care should be taken in directing the axis of the coil. Figure 4.12 indicates how the eddy current paths are diverted by the presence of a surface crack. Using purely analytical methods, attempts have been made to predict how these changes relate to the size and shape of the defect but with only a very limited degree of success (Dodd, 1977). More realistic results have been obtained by the following modelling techniques:

- Mechanically by introducing artificial defects either in the form of cuts in solid metals or of solid inclusions in liquid metals.
- Theoretically by the use of numerical methods.

Whichever method is used, with the concept of normalization, a single set of results can be adapted for a particular testing arrangement for any metal of any coil size and any frequency, with certain constraints,

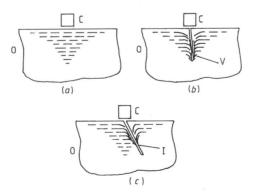

**Figure 4.12** Vertical cross-section of eddy current flow patterns (broken lines) in a conductor (O) beneath a surface-scanning coil (C) for (a) no defect, (b) a vertical crack (V) and (c) an inclined crack (I).

such as a fixed diameter ratio for tube testing with an internal axial coil and a given normalized length for a coil scanning a surface.

Although all of the coil arrangements which have been discussed are capable of detecting whether a defect is present within a given region, it is only the surface-scanning technique that can be used to locate its precise position. With rod and tube testing, for example, an encircling coil or an internal coil method can be used initially to provide a rapid overall scan of the sample and, if any adverse indications are observed, the surface-scanning technique with a much smaller coil is then used to provide a more detailed examination of the relevant part of the test object.

With encircling coils, the parameter for crack sizing (Figure 4.13) is depth $d$ of the crack tip below the surface, normalized by dividing by $b$, i.e. the radius for a rod or the external radius for a tube, to give the normalized value $d_0 = d/b$. Förster preferred to divide $d$ by the diameter for this purpose. The thickness $t$ of the crack can be normalized in the same way, i.e. $t_0 = t/b$, or just expressed as a ratio to the depth since $t/d = t_0/d_0$. We use here the term 'width' $w$ to describe the distance along the crack in the downward direction, to avoid confusion with length, which is taken here to indicate longitudinal dimension in the plane of the surface. When a crack is inclined at an angle $\theta$ to the normal to the surface, $w \cos\theta = d$ and $w = d$ for a surface-breaking crack perpendicular to the surface. The effective radius $r_0$ replaces $b$ in the above expressions when a surface-scanning coil is used.

The size of a surface-breaking crack, for example, can be assessed by moving a probe coil from a position on the test object where there is no defect to a position where a crack is present, measuring the resultant changes in the impedance components of the coil. These changes depend on the depth $d$ and the thickness $t$ of the crack but it is the magnitude of $d$ which determines the maximum impedance change. Figure 4.12b clearly indicates this because any increase in the value of $t$ only increases the area over which the eddy currents are disturbed, hence the area over which the change of impedance is observed. For a given coil/sample arrangement, the value of this impedance change

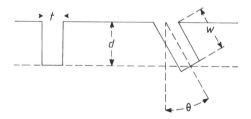

**Figure 4.13** Dimensions used to describe normal and inclined surface cracks: $d$ = depth, $t$ = thickness, $w$ = width, $\theta$ = angle of inclination.

## Defect modelling

also depends on the normalized values of frequency and lift-off (or fill-factor), the radius of the coil (for surface scanning) and the ratio of the internal diameter to the external diameter (for tube testing with either encircling or internal axial coils).

Maximum resolution of the 'defect signal', i.e. the vector representing the changes $\delta(\omega L)$ and $\delta R$ in the two components of impedance, occurs at some optimum frequency for a given electrical conductivity $\sigma$ and, for non-saturated ferromagnetic conductors, at a given magnetic permeability $\mu$. The corresponding optimum normalized frequency $f_0$, where $f_0 = \omega \mu \pi r_0^2$, can then be calculated. This is a relatively simple matter for non-ferromagnetic or saturated ferromagnetic materials, but difficulties occur with non-saturated ferromagnetics because of uncertainties arising when defining the appropriate value of the magnetic permeability (section 2.7.1).

The optimum value of $f_0$ for testing with a coil of radius $r_0$ scanning the surface of a non-ferromagnetic metal having a known value of $\sigma$, can be obtained with the aid of a test block of this metal containing saw-cuts simulating cracks of different known depths $d$. The scanning is performed with a given normalized value of lift-off $k$, where $k = h/r_0$, which is usually the minimum possible value, and the curve relating impedance to crack depth, i.e. the defect curve, is obtained. The frequency is varied until the phase difference between the defect curve and the $\delta\sigma$ (i.e. $\delta f_0$) curve is a maximum (e.g. Figure 4.14). The calculated corresponding value of $f_0$ is then the

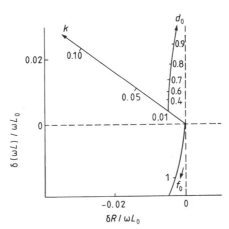

**Figure 4.14** Variation of the normalized components of impedance in a coil scanning across a simulated vertical surface crack in a non-ferromagnetic conductor. The curves are obtained from values for a defect-free region at normalized frequency $f_0 = 63.3$ and normalized coil length $k_0 = 1.5$. The parameters $d_0$ and $k$ are respectively the normalized crack depth and the normalized lift-off. (Reprinted, by permission, from Hajian and Blitz, 1986; copyright Butterworth-Heinemann Ltd)

optimum normalized frequency which can be used for testing any other non-ferromagnetic metal with the particular coil used and with the same value of $k$, provided there is a suitable impedance match with the AC bridge (section 5.2.1).

### 4.6.2 Mechanical modelling

The reference to saw-cuts having varying depths in blocks made from the material of the sample to be examined is everyday practice when calibrating eddy current equipment for defect sizing. Difficulties arise because it is very rare for a real crack to bear much resemblance to a uniformly produced saw-cut; the saw-cut may well be much thicker than a real crack, although considerable improvements have been achieved with the use of electron beams to provide very thin cuts, e.g. 0.2 mm. Nevertheless, the use of saw-cuts is usually of great value because it provides a very good indication of the depth of a crack tip below the surface. It can be easily verified that the maximum changes of impedance of a coil scanning a surface are the same for both cracks and saw-cuts having similar depths, irrespective of thickness or angle of inclination. The shape, orientation and thickness of a crack determine how the impedance changes as the probe moves across it.

A great advantage of the use of mechanical models is the ability to scale up dimensions, for a given value of normalized frequency, so as to enable more realistic relative crack sizes to be attained.

Förster (1986) is on record as being the first to carry out defect modelling in a systematic manner. This was done with coils encircling, with maximum possible fill-factor, cylindrical mercury bars containing simulated surface-breaking and subsurface longitudinal cracks, oriented in radial directions, and also spherical voids. The metal cylinders were formed by pouring mercury into a glass tube and the defects were simulated by firmly locating pieces of plastic material in the desired positions. Impedance measurements were made for simulated defects of different sizes at a number of frequencies and useful results were obtained for normalized frequencies $f_0$ ranging from 5 to 150. At higher values of $f_0$ the sensitivity of detection decreased rapidly and at lower values the phase angle between the $\delta D$ curve and the 'defect' curve became too small for accurate resolution. Figure 4.15 shows the variation between crack depth and impedance for various depths of the upper ends of cracks below the surface, indicated as percentages of tube diameter for $f_0 = 15$, the optimum normalized frequency. Förster found that the impedance decreased as the crack thickness was reduced to a lower limit of 1% of tube diameter, below which no further change of impedance could be detected.

# Defect modelling

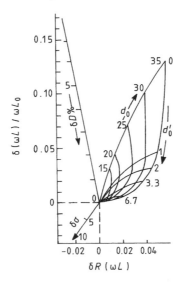

**Figure 4.15** Variation of the normalized impedance components $\delta(\omega L)/\omega L_0$ and $\delta R/\omega L_0$ for a coil encircling a non-ferromagnetic conducting cylindrical rod containing defects, plotted with depths of radial cracks directed longitudinally for a normalized frequency obtained by Förster (1986) using mercury models. The parameter $d_0$ is the crack depth and $d_0'$ is the depth of the upper end of a subsurface crack; both are expressed as a percentage of the cylinder's diameter $D$. Variations of normalized impedance components with $D$ and electrical conductivity $\sigma$ are also shown. (Reprinted by permission of the American Society for Nondestructive Testing)

The electrical conductivity of mercury is 1.05 MS m$^{-1}$, compared with values ranging from about 5 to over 60 MS m$^{-1}$ for most commonly used non-ferromagnetic metals, although Inconel 600, an alloy used in the manufacture of steam generator tubes, does have a conductivity very close to that of mercury (Junker and Mott, 1988). This permits the use of larger dimensions for models because a decrease in $\sigma$ is compensated by an increase of the square of the rod radius $b$. To apply the results obtained from mercury cylinder modelling to predicting coil impedances with a sample of any other non-ferromagnetic material, e.g. copper or aluminium, irrespective of its diameter but for a given fill-factor, it is necessary to calculate the operating frequency corresponding to the normalized frequency $f_0$ ($f_0 = \omega \mu \sigma b^2$) used for the mercury model. If the radius in the model is 25 mm, the frequency corresponding to Förster's optimum value of $f_0 = 15$ is 2.9 kHz and for an aluminium alloy cylindrical rod of radius 4 mm and an electrical conductivity of 25 MS m$^{-1}$ the corresponding operating frequency is 4.8 kHz. As an example, a simulated radial surface-breaking crack, depth 10% of the rod diameter, is 5 mm deep in the mercury model tested at the optimum frequency of 2.9 kHz and 0.8 mm deep in the alloy rod tested at 4.8 kHz. The change in normalized impedance when the defect is introduced is the same in both instances.

Förster also modelled radial defects in mercury tubes and further work in this field has been carried out by Aldeen and Blitz (1979) for cracks inclined to the radial direction. More recently, measurements have been made with surface-scanning coils by Junker and Mott (1988) on mercury-simulated plates.

There are several reasons why mercury is not a highly satisfactory material for defect modelling:

- It is expensive.
- It is highly toxic and requires strict precautions in its handling.
- It forms a convex meniscus which is difficult to suppress and thus leads to widening of the simulated cracks at the surface.
- The finite thickness of the glass containing the mercury limits the value of the fill-factor.

On the other hand, the use of mercury provides an excellent means of simulating internal defects, for which there are no menisci to create problems.

Some of the difficulties encountered with mercury models have been overcome with the use of Wood's metal (Blitz and Alagoa, 1985), an alloy which is solid at room temperatures, melts at about 70 °C and is relatively inexpensive. Its use is particularly effective for surface-scanning measurements. Wood's metal has an electrical conductivity of about 1.9 MS m$^{-1}$ at room temperature, about twice that of mercury; it is thus more practical for the scaling of dimensions. A rectangular mould of suitable size is constructed from say aluminium plates, and slots are cut into two parallel walls. Simulated defects usually made from thin mica sheets of the required sizes are located in these slots so as to be either perpendicular or oblique, at a given angle, to the surface of the Wood's metal. The molten metal is then carefully poured into the mould, so as to avoid air bubbles. After cooling, the metal, now a solid block, is removed and the upper surface machined flat so that impedance measurements can be performed with a surface-scanning coil of suitable dimensions at a given degree of lift-off.

The dimensions of the block used by the author and his colleague were 90 mm × 90 mm × 30 mm. The thickness of 30 mm was sufficient to prevent through-penetration of eddy currents to the lower surface, taking into account the penetration depth $\delta = 2.6$ mm for a frequency of 20 kHz. (It is readily seen that $\delta^2 = 2r_0^2/f_0$.) Cracks were simulated by flat mica strips, 0.1 mm thick, of varying widths and extending over the width of the block.

Consider, for example, an aluminium alloy test block ($\sigma = 19.2$ MS m$^{-1}$) containing saw-cuts and scanned with a constant lift-off $h$ of 0.02 mm (normalized value $k = h/r_0 = 0.001$) with an air-cored coil of radius $r_0 = 2$ mm and length $l = 3$ mm ($k_0 = l/r_0 = 1.5$). The maximum resolution between the impedance vectors representing crack size and $\delta f_0$ occurs at a frequency of 100 kHz (i.e. for $f_0 = 63.3$). For the same value of $f_0$ with a Wood's metal block ($\sigma = 1.9$ MS m$^{-1}$) tested at a frequency of 20 kHz, $r_0$ and $l$ are equal to 14 mm and 21 mm, respectively, thus producing a scale factor of 7. Thus, $h$ becomes equal to 0.07 mm and a 1 mm deep crack in the alloy is equivalent to a crack 7 mm deep in the Wood's metal A thickness of 0.1 mm of the simulated crack in Wood's metal is equivalent to one of 0.014 mm in the aluminium alloy, the same order of magnitude as for a real crack.

The changes in the components of impedance of the coil as it scans across the simulated crack are indicated by a loop on the impedance diagram (Figure 4.16). Considering a number of plane cracks, each having the same depth $d$ below the surface, i.e. $d = w/\cos\theta$, the area of the loop increases with the angle $\theta$ of orientation, commencing with a curved line when $\theta = 0$. This phenomenon can easily be explained with reference to Figure 4.12c, which illustrates a vertical section of the eddy current lines of flow and thus indicates a greater departure from symmetry with increasing angle $\theta$.

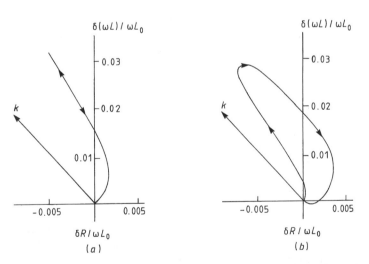

**Figure 4.16** Variation of the normalized components of impedance in a coil scanning the surface of a conductor for (a) a perpendicular crack and (b) a crack inclined at 50° to the surface normal (cf. Figure 4.12). The curves were obtained for Wood's metal by Blitz and Alagoa. Arrows indicate the direction of scanning. Both cracks were 5 mm deep, the coil radius was 5 mm and the operating frequency was 20 kHz, providing a normalized crack depth $d_0 = 1$ and a normalized frequency of 7.6.

128                    *Eddy current principles*

### 4.6.3 Numerical modelling

Some early work on modelling defects has been carried out using the finite difference method (Ida, 1986b) but the more effective finite element method is now commonly practised. In the limited amount of space available here it would be impossible to present an adequate account of the principles of this method and the interested reader is referred either to Rao (1989) or Ida (1995).

With the finite-element method, the region of interest, which includes the coil, the object to be tested and any intervening space is divided into elementary areas or volumes, i.e. finite elements, depending on whether two or three dimensions are considered, and each of them possesses a given value of electrical conductivity $\sigma$ and magnetic permeability $\mu$. For convenience, the shapes of the elements are kept as simple as possible and often take forms such as rectangles, triangles, rectangular blocks and tetrahedra, with either straight or curvilinear boundaries. One side or surface of an element may coincide with a discontinuity such as the boundary of a crack or a surface of the material being tested. Thus, within a defect-free region of a homogeneous metal, each element has the same finite values of $\sigma$ and $\mu$, but for elements inside an air-filled crack, the values of $\sigma$ and $\mu$ are equal to zero and $\mu_0$, respectively. Equation (4.40) can be applied to compute values of the magnetic vector potential $A$ at each element in the system. To minimize the number of computations, the sizes of these elements are made to vary in relation to the degree of precision required, i.e. they are large where variations of $A$ are small and small where they are large.

Much effort is saved with an axisymmetric system, as used in the example given below, because only two dimensions need to considered. But with most applications, such as the sizing of surface cracks with a scanning coil having its axis perpendicular to the surface, asymmetric geometries occur and the use of a three-dimensional mesh of elements is necessary.

A good understanding of the application of finite-element analysis can be obtained from the work of Lord and Palanisamy (1981), who used a pair of differential internal axial coils to test a tube with a uniform slot cut into its external wall (Figure 4.17). Because the system is axisymmetric, only two dimensions are required. With this analysis, an array of elements is set up in the $r-z$ plane, each having a thickness. The elements are triangular in shape except within the coils, where they are rectangular, and the non-linear energy function $F$ is related to the magnetic vector potential $A$, flux density $B$ and current density $J$ (equation 4.40) as follows:

$$F = \iiint \left( \int (1/\mu) \mathbf{B} \cdot d\mathbf{B} + (1/2) j\omega\sigma |A|^2 - \mathbf{J} \cdot \mathbf{A} \right) dV \qquad (4.68)$$

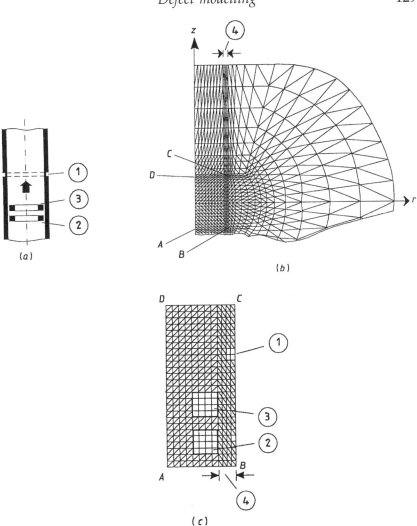

**Figure 4.17** Eddy current testing of a non-ferromagnetic conducting tube with an annular external slot using internal differential coils: (a) arrangement of coils and tube; (b) mesh structure in the $r-z$ plane for the entire region of interest; (c) magnified mesh structure *ABCD* in the region of the coils and the slot. 1 = defect, 2, 3 = coils, 4 = tube wall. (Reprinted, by permission, from Lord and Palanisamy, 1981; copyright the American Society for Testing and Materials)

where d$V$ is the elementary volume. As before, **A**, **B** and **J** vary as exp(j$\omega$t) and are functions only of $r$ and $z$, assuming axisymmetry. $F$ is minimized by setting $\partial F/\partial A_k = 0$, where $k = 1, 2, \ldots N$, and $k$ indicates the vertex value and $N$ the total number of nodes in the region under consideration. The equations for each element are combined

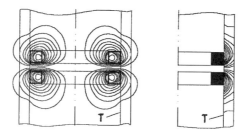

**Figure 4.18** Contours of (a) magnetic vector potential amplitude in the region of the tube remote from the slot and (b) eddy current density amplitude in the material of the tube. T = the tube wall. (Reprinted, by permission, from Lord and Palanisamy, 1981; copyright the American Society for Testing and Materials)

into a single 'global matrix' equation and the impedance $Z_i$ for each filament of the coil, of which there may be several per turn, is

$$Z_i = -j\omega 2\pi r_i A_i \qquad (4.69)$$

Performing integrations produces values for the impedances $Z_a$ and $Z_b$ of the two coils wound in opposition and the resultant impedance $Z$ of the probe is equal to the vector sum of these two quantities, i.e.

$$Z = Z_a + Z_b \qquad (4.70)$$

Figure 4.18 shows contours for the amplitude of $A$ within the system and for eddy current density within the wall of the tube in a defect-free portion remote from the annular slot. Figure 4.19 illustrates the impedance loop obtained for a scan of the probe.

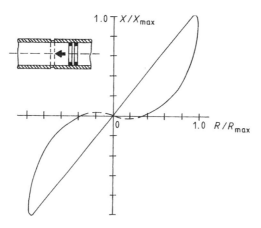

**Figure 4.19** Prediction of changes in the components of impedance $X$ and $R$ for the differential coil probe, changes due to the presence of the slot during the scan. $X$ and $R$ are normalized with respect to their maximum values $X_{max}$ and $R_{max}$. (Reprinted, by permission, from Lord and Palanisamy, 1981; copyright the American Society for Testing and Materials)

A non-mathematical account of numerical modelling techniques for eddy current and other electrical and magnetic testing methods has been given by Becker *et al.*, (1986).

# CHAPTER 5

# Eddy current methods

## 5.1 GENERAL CONSIDERATIONS

The scope of eddy current testing is summarized in section 4.1. But with few exceptions the use of eddy currents is restricted to testing samples made either wholly or partly from metal and then only for regions at or just below the surface. However, the technique has the distinct advantage that no contact with the object under test is required and that surface preparation is usually not necessary. As stated in section 4.1, the method consists of passing an alternating current through a coil so as to excite coaxially circular currents, i.e. eddy currents, in an electrically conducting object in its vicinity. The resulting changes in the components of the impedance of the coil are characteristic of the electrical and magnetic properties of that part of the object in which the eddy currents are flowing. A single coil is generally used but two coils are sometimes employed (see later); one coil excites the eddy currents and the other coil detects them. The essential parts of eddy current equipment are the exciting oscillator, the measuring circuit, usually an AC inductance bridge, and the detecting coil (or coils). The object to be tested can be scanned either manually or with the aid of a mechanical device. In mechanical scanning the signals denoting the positions of the coil are often synchronized with the output signals from the detector.

The treatment in this chapter covers the techniques and applications of eddy current testing. The basic methods are dealt with first of all. More recent developments, including some applications which, until quite recently, were generally considered outside the scope of eddy current testing are discussed later and in Chapter 6.

## 5.2 FUNDAMENTAL MEASUREMENTS

### 5.2.1 Basic principles

Eddy current testing requires us to determine the components of the impedance of the detecting coil or the potential difference across it.

Most applications require the determination only of changes in impedance, which can be measured with a high degree of sensitivity using an AC bridge. The principles of operation of the most commonly used eddy current instruments are based on Maxwell's inductance bridge (section 2.6.2) in which the components, i.e. $R_3$ and $\omega L_3$, of the impedance of the detecting coil, commonly called a probe, are compared with known variable impedances $R_4$ and $\omega L_4$ connected in series and forming the balancing arm of the bridge (Figure 2.12b). If $R_1$ and $R_2$ are the resistances of the ratio arms, the conditions for balance are given by

$$R_1/R_2 = R_3/R_4 = L_3/L_4 \tag{5.1}$$

where the values of $R_1$ and $R_2$ are generally fixed and equal to one another, and $L_4$ and $R_4$ are variable.

The input EMF to the bridge is an AC oscillator, often variable in both frequency and amplitude. The detector arm takes the form of either a meter or a storage cathode-ray oscilloscope, a phase-sensitive detector, a rectifier to provide a steady indication and usually an attenuator to confine the output indication within a convenient range. Storage facilities are necessary in the oscilloscope in order to retain the signal from the detector for reference during scanning with the probe.

An alternative display is an ellipse appearing on the screen of a non-storage type cathode-ray oscilloscope as used with Förster's Sigmaflux instrument for testing tubes with an encircling coil (McMaster, 1963d, Blitz, et al., 1969a). As far as the author is aware, the Sigmaflux with its straightforward design is no longer being manufactured but is still being used. The out-of-phase components of the output potential difference are left unrectified and are fed separately, after suitable amplification, to the X- and Y-plates of the oscilloscope so as to provide the elliptical display. The orientation of the major axis of the ellipse and the value of its axis ratio indicate, respectively, the phase and amplitude of the output signal. On inserting a defect-free sample, a straight line appears on the oscilloscope screen. The reference phase angle is then varied until this line becomes horizontal. When the defect-free sample is replaced by a defective sample, the ellipse is observed. A calibration can be made by using test samples containing simulated defects and the characteristics of the ellipse related to the location and size of any defect present.

Whichever method is used, the highest sensitivity of detection is achieved by properly matching the impedance of the probe to the impedance of the measuring instrument. Thus, with a bridge circuit (Figure 2.12) which is initially balanced, a subsequent but usually small variation $\delta Z_3$ in the impedance $Z_3$ of the probe upsets the balance, and a potential difference $\delta V$ appears across the detector arm of the bridge. The sensitivity of measurement can be expressed as the ratio $\delta V/\delta Z_3$, and by applying Kirchhoff's laws to the bridge circuit, it can easily be shown how this ratio is a maximum when the impedances $Z_1$, $Z_2$, $Z_3$ and $Z_4$ are equal to one another (e.g. Harnwell, 1938). Usually $Z_1$ and $Z_2$ are purely resistive, whereas $Z_3$ and $Z_4$ are both complex quantities and therefore frequency dependent. Hence, for large changes in frequency, adjustments are required in the values of $R_1$ and $R_2$. Most variable-frequency instruments operate over several different bands and, when the frequency band is changed, the values of $Z_3$ and $Z_4$ change in an abrupt manner but, at the same time, different values of $R_1$ and $R_2$ are automatically switched into the circuit to maintain optimum conditions for matching.

Although the Maxwell inductance bridge forms the basis of most eddy current instruments, there are several reasons why it cannot be used in its simplest form (e.g. Hague, 1934), including the creation of stray capacitances, such as those formed by the leads and leakages to earth. These unwanted impedances can be eliminated by earthing devices and the addition of suitable impedances to produce one or more wide-band frequency (i.e. low $Q$) resonance circuits. Instruments having a wide frequency range, e.g. from 1 kHz to 2 MHz, may possess around five of these bands to cover the range. The value of the impedance of the probe is therefore an important consideration in achieving proper matching and, as a result, it may be necessary to change the probe when switching from one frequency band to another.

### 5.2.2 Simple calibration

The instrument may be calibrated for a particular application by balancing the bridge when the probe is applied to a standard test object which has known physical properties. When a surface-scanning coil is used to measure depths of surface cracks in metals, the standard takes the form of a test block made from the same metal as the samples to be tested and containing a series of saw-cuts having different depths. The bridge is balanced with the probe located over a defect-free portion of the test block with zero or a small constant value of lift-off (section 4.6.1). On moving it to one of the saw-cuts, there are changes $\delta R$ and $\omega \delta L$ in the impedance components of the probe and the bridge becomes unbalanced. Rather than rebalancing the bridge to evaluate $\delta R$ and $\omega \delta L$, generally small compared with $R_3$ and $L_3$, it is often more expedient to note the amplitude $|\delta V|$ of the output potential difference across the detector, usually called the output signal, and the phase angle $\phi$ with respect to some reference, e.g. the input EMF. This procedure is repeated for the other saw-cuts and a calibration curve is plotted. The calibration curve can only be used for a given frequency and for a fixed value of the input EMF.

It is usually more convenient to use, as a phase reference, the change in impedance vector (or the corresponding output signal from the detector) arising from variations in either lift-off or fill-factor (sections 4.2 to 4.5). This vector, initially linear, can be checked readily at any time during a test. By observing only those changes in impedance (or output signal) having a phase difference of 90° with respect to the reference vector, it can be ascertained whether there are no unwanted contributions to the output voltage. If the detector has an oscilloscope output, the phase control is repeatedly adjusted and, at the same time, the probe is raised and lowered (or, with an encircling coil, the test object is slid in and out) until the initially linear portion of the unwanted vector (Figures 4.2 to 4.5) is displayed as a horizontal line; the position of the line is then moved to coincide with a suitably located horizontal line on the graticule in front of the screen. The measurements of the required parameter, e.g. crack depth, electrical conductivity, are then obtained simply by observing the vertical displacement of the signal above this line.

### 5.2.3 Calibration for impedance changes

A more fundamental method, independent of the value of the input EMF and the characteristics of any particular instrument containing an inductance bridge, is to perform calibration directly with respect to the changes in the impedance components of a given coil at some fixed frequency. Assuming the measured impedance changes $\omega \delta L$ and $\delta R$ of the probe are small compared with the balanced impedance components $\omega L_3$ and $R_3$, it can be shown (Blitz et al., 1981) that

$$V_R = A\delta R \text{ and } V_L = A\omega\delta L \qquad (5.2a)$$

$$V_L/V_R = \omega\ \delta L/\delta R = \tan \phi \qquad (5.2b)$$

where $V_R$ and $V_L$ are the components of the output potential difference $\delta V$ across the detector resulting from the imbalance of the bridge brought about by changes in the impedance of the probe, i.e. for a given frequency, $V_R$ and $V_L$ are resolved in the directions of the variations of $R_3$ and $L_3$, respectively. $A$ is a constant having a value that depends on the characteristics and settings of the AC bridge and its associated instrumentation and on the frequency of the applied EMF. For most applications, the linearities assumed in equation (5.2a) are correct to about 1%.

The values of $A$, hence the phase angle $\phi$, can be obtained from a simple calibration using a small resistor of known value, typically 0.2 $\Omega$, with a switch connected in parallel. This arrangement is inserted in series with the probe. The instrument is initially balanced with the switch in the 'on position' and short-circuits the series resistor. The switch is moved to the 'off position' so the resistor is in series with the probe coil and the amplitude is measured for the output potential difference $V_R$ across the detector, which depends only on the change in resistance; the constant $A$ (equation 5.2a) can then be evaluated. The term $V_R$ then indicates values of $\delta R$ and its direction defines the zero value of $\phi$. With an oscilloscope output, the phase-sensitive detector can be rotated until the displaced signal representing $V_R$ is horizontal; equation (5.2a) is then used to calibrate the horizontal scale in ohms. The direction of the inductive component vector $\delta(\omega L)$ is clearly vertical and the magnitude is again obtained from equation (5.2a); this has been confirmed experimentally.

To relate, experimentally, the measured impedance changes to the normalized values used to plot the impedance analysis curves (Figures 4.2, 4.3, etc.), it is necessary first to measure the absolute values $R_3$ and $\omega L_3$ of the components of impedance of the probe for the different values of frequency and either lift-off or fill-factor when located at a defect-free region of the reference sample. The size of $R_3$ is evaluated by subtracting the ohmic resistance of the windings of the coil from the measured resistance. In addition, the absolute value of the inductive impedance $\omega L_0$ of the probe when remote from the test sample should be measured to enable normalization to be achieved, i.e. by dividing it into $\omega L_3$ and $R_3$. These procedures may be difficult to carry out in practice with commercial eddy current instruments, which are mostly designed to indicate only *changes* in impedance. Hence a high-quality AC bridge is almost certainly required.

Recent advances in microprocessing have led to the production of eddy current instruments which can perform automatic calibrations for direct readings of impedance components and, consequently, of defect sizes, electrical conductivity, dimensions, hardness values, etc.

### 5.2.4 Resonance bridge method for crack detection

Cecco *et al.* (1981) describe a resonance bridge circuit which is suitable for the design of the portable eddy current crack detector and provides a moderate degree of precision. Three of the arms of the bridge (Figure 2.12a) are purely resistive, i.e. $\mathbf{Z}_1 = R_1$, $\mathbf{Z}_2 = R_2$ and $\mathbf{Z}_4 = R_4$ where $R_1$ and $R_2$ are fixed and $R_4$ is variable. The probe coil, having an inductance $L$ and a resistance $R_1$ has a variable capitance $C$ shunted across it to form the arm $\mathbf{Z}_3$ and the value of $C$ is adjusted until the coil resonates, i.e. when $\omega L = 1/\omega C$. Provided $R$ is very much less than $\omega L$, $\mathbf{Z}_3$ becomes a pure resistance and equal to $1/R$ (Buckingham and Price, 1966). Balance is achieved when

$$L/CR = R_1 R_4 / R_2$$

If the coil is located in a defect-free region of the test sample, where the bridge is balanced, then moved to the position of a defect (section 5.2.2), the bridge is no longer balanced and the out-of-balance voltage output of the detector can be related to the size of the defect as a result of the consequent changes in $L$ and $R$.

## 5.3 PROBE DESIGN

The most important feature in eddy current testing is the way in which the eddy currents are induced and detected in the material under test. This depends on the design of the probe, which can contain either one or more coils. A coil consists of a length of wire wound in a helical manner around the length of a cylindrical tube or rod, called a former. The winding usually has more than one layer so as to increase the value of inductance for a given length of coil. It is desirable with eddy current testing that the wire is made from copper or other non-ferrous metal to avoid magnetic hysteresis effects. The main purpose of the former is to provide a sufficient amount of rigidity in the coil to prevent distortion. Formers used for coils with diameters greater than a few millimetres, e.g. encircling and pancake coils, generally take the form of tubes or rings made from dielectric materials. The region inside the former is called the core, which can consist of either a solid material or just air. Small-diameter coils are usually wound directly on to a solid core, which acts as the former.

The higher the inductance $L$ of a coil, at a given frequency, the greater the sensitivity of eddy current testing. A more precise value of $L$ (cf. equation 2.49) is given by

$$L = Kn^2\pi[(r_0^2 - r_c^2) - \mu_r r_c^2]\mu_0/l \tag{5.3}$$

where $r_0$ is the mean radius of the coil, $r_c$ the radius of the core, $l$ the length of the coil, $n$ the number of turns, $\mu_r$ the relative magnetic permeability of the core, $\mu_0 = 4\pi \times 10^{-7}$ H m$^{-1}$ (i.e. the permeability of free space which is effectively equal to the permeabilities of the materials of both the wire and the former) and $K$ a dimensionless constant characteristic of the length and the external and internal radii.

It is essential that the current through the coil is as low as possible. Too high a current may produce (a) a rise in temperature, hence an expansion of the coil, which increases the value of $L$, (b) magnetic hysteresis which is small but detectable when a ferrite core is used and (c) for ferromagnetic materials, excessive magnetic hysteresis accompanied by non-linearity of the output signal, which leads to the appearance of harmonic frequencies.

The simplest type of probe is the single-coil probe, which is in widespread use. Sometimes it is desirable to use a probe consisting of two coils arranged in transformer fashion and therefore known as a transformer probe (Figure 5.1). Here the primary coil induces eddy currents in the test object and the secondary coil acts as a detector. The use of this probe provides an enhanced signal-to-noise ratio for detection, advantageous when deep penetration is required, e.g. for seeking internal defects.

The through-transmission method is sometimes used when complete penetration of plates and tube walls is required; in these cases a fork probe may be used. With a fork probe, the coils are held in position by a calliper which locates them a fixed distance apart on opposite sides of the relevant section of the test sample. Figure 5.1d illustrates a fork probe design consisting of a transmission coil with the two receiving coils separated by the object. The relationships between the impedance components of the coils characterize the properties of the intervening metal plate (Förster and Libby, 1986). The three basic probe positions (Figure 4.1) are

- Encircling: for the external testing of cylindrical tubes and rods.
- Internal axial: for the internal testing of cylindrical tubes.
- Surface scanning: where the axis of the coil is at right angles to the surface, including the inner surface of a tube.

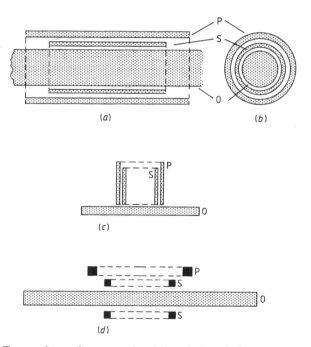

**Figure 5.1** Types of transformer probe: (a) encircling (side view), (b) encircling (end view), (c) surface scanning, (d) through-transmission (fork probe). P = primary, S = secondary, O = object.

The choice of the term 'internal axial' is the author's. The American Society of Materials and Testing (ASTM E 268-84 a: Definition of terms relating to electromagnetic testing) offer a selection from four different names, i.e. ID (internal diameter) coil, inserted coil, bobbin coil and inside coil, none of which indicates that the direction of the axis of the coil is the same as for the tube. Encircling and internal axial probes allow rapid scanning with a single sweep and are useful for measuring the bulk properties of homogeneous samples, i.e. electrical conductivity, magnetic permeability and dimensions. Surface-scanning probes are normally used for the localized assessment of discontinuities such as defects and other structural variations. The radius of a surface-scanning coil should be sufficiently small for the area directly underneath it to be effectively plane, and the optimum radius depends on the degree of curvature of the scanned surface and the required degree of sensitivity of detection, taking into account that a reduction in radius of the coil leads to a lower value of inductance (equation 5.3).

The radii of encircling and internal axial coils are determined by the radii of the tubes and rods under test but the radii of surface-scanning coils depend on the type of application. A high degree of resolution is usually necessary for flaw detection and radii of 1 or 2 mm are common, but for other applications, e.g. conductivity measurements of homogeneous metals, larger-diameter coils are often used.

The speed and precision of scanning can be improved with the use of a mechanical device and, for tubes and rods, the test sample can be rotated as it moves in an axial direction so as to provide a helical scan. With surface scanning, the probe is fixed in a given position which allows a constant and small amount of lift-off. The scanning speed can be increased by using an array of probes that are suitably spaced to allow for complete surface coverage. Ida (1986a) has remarked that the eddy current field in the sample extends to a distance from the axis of three coil diameters.

The core of an encircling coil, when in the testing position, is clearly the object being examined and the value of its magnetic permeability depends on the nature of its material. For the other types of coil, the material of the core is a matter of choice. Because the magnetic permeabilities of dielectric materials are virtually equal to the permeability of air, i.e. $\mu_0$, coils with dielectric cores are often described as being air-cored. However, equation (5.3) shows that the value of $L$ can be raised considerably by using a core with a high magnetic permeability. Ferrites are highly suitable materials for this purpose. With small exciting currents, an increase of from five- to tenfold in the value of $L$ is possible, depending on the value of the recoil magnetic permeability (section 2.7.1), when a ferrite core is inserted into a coil. Ferrites have electrical conductivities of the order of $100\,S\,m^{-1}$, i.e. a fraction of about $10^{-5}$ of typical values for metals, and the amount of eddy current induction in them is negligible for small coil currents. The use of ferrite-cored coils is advantageous for testing non-ferromagnetic and saturated ferromagnetic metals but difficulties arise when testing unsaturated ferromagnetic metals because the eddy currents induced in the test material by the coil give rise to a magnetic field. This field may be sufficiently high to produce a detectable amount of hysteresis and parasitic eddy currents in the core through consequent changes in the magnetic permeability of the test sample. Consequently, the impedance of the coil is no longer constant, even with small varying currents and the use of ferrite cores is not recommended for measurements on non-saturated ferromagnetic metals. However, ferrite-cored probes are highly effective for detecting the existence of cracks in steels which, in some instances, may be difficult to find by other methods.

Although equation (5.3) shows how the use of longer coils produces a higher value of $L$, there is an effective upper limit to the length because the regions of a coil remote from the surface have little effect on the induced eddy currents, and vice versa.

The value of $L$ can also be increased by enlarging the area of cross-section $A$, but this reduces the resolution for detection of small discontinuities. A compromise can be reached by using multilayer windings; multilayer windings achieve a corresponding decrease in length without too great an increase in the effective value of $A$, yet they provide a high degree of sensitivity. A coil designed in this way is called a **pancake coil** (section 4.5.1); pancake coils are advantageous for measuring the electrical conductivity of a homogeneous metal. Although it may not be highly satisfactory for resolving discontinuities which are close together, a pancake-coil probe is very suitable for sizing defects which are remote from other discontinuities, e.g. from other defects and from edges.

Increasing the sensitivity of eddy current measurements can also be obtained by using a shielded probe which restricts the spreading of the lines of magnetic flux away from the region of interest. This type of probe has the advantage of reducing the possibility of unwanted signals being received from neighbouring discontinuities such as edges and bolt holes. The cup-core (or pot) probe (Figure 5.2) has also proven effective for the surface-scanning of non-ferromagnetic materials. The core and the shielding cup, made from ferrite, form a single unit and provide a low magnetic reluctance (section 2.7.2). The emerging flux is concentrated in a region immediately below the probe. Ida (1986b) discusses the results of a numerical modelling method applied to a cup-core probe which shows how the normal component of the magnetic field just below the lower end of the coil is twice that for a ferrite-cored probe and 10 times that for an air-cored

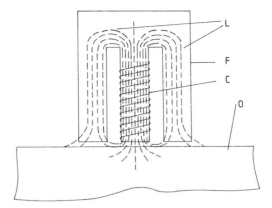

**Figure 5.2** Cup-core or pot (probe). C = coil winding, F = ferrite core or cup, O = object under test, L = magnetic flux lines.

probe. In each instance the coils have the same size and the current and frequency (100 kHz) are constant. Ida also describes the use of copper-shielded probes for localizing electromagnetic fields when testing ferromagnetic materials.

Wincheski *et al.* (1994) have used what they call a **self-nulling** eddy current probe for surface and subsurface flaw detection. It consists of a transformer probe, typically 12.7 mm in diameter (Figure 5.3) with the primary coil (P) and the secondary coil (S) completely isolated magnetically from one another so as to avoid any direct flux linkage between them. This is achieved by covering the secondary coil with a ferromagnetic cup (F). An alternating EMF of constant amplitude and frequency is applied to the input of the primary coil and a magnetic flux is applied so as to induce eddy currents in the conducting test sample (M), thus giving rise to a magnetic flux passing through the secondary coil.

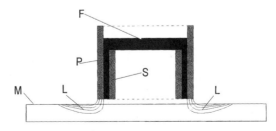

**Figure 5.3** Basic design of the self-nulling probe: P = primary coil, S = secondary coil, M = metal test sample, F = ferromagnetic shield, L = eddy current flow lines. (Reprinted from Wincheski *et al.*, 1994, by permission of the American Society for Nondestructive Testing)

The sample is scanned at a constant speed, and in the absence of cracks or other discontinuities, eddy currents having fixed circular shapes appear in the sample (Figure 5.4a) and a magnetic flux having a fixed value is induced in the secondary coil. The alternating potential difference is measured across the output of the secondary coil, in accordance with Faraday's law (equation 2.18). If a crack or other discontinuity is present in the sample, the circular patterns of the eddy currents are disturbed (Figure 5.4b and c) and changes take place in the value of the induced magnetic flux, thus creating a potential difference across the output terminals of the secondary coil. The value of this potential difference can be related to the size and extent of the defect.

An advantage found with the self-nulling probe is there is no output from it when a defect is not present and impedance balancing is unnecessary. Together with its comparatively small size, this provides an increase in the speed of inspection. The designers have been able to use it to measure isolated defects as small as 2.5 mm, sometimes even smaller, and they have shown how the device can prove useful in detecting notches at rivets and corrosion in metals. They have also shown that, because the voltage output produces only a distortion of the eddy current pattern, this probe is less sensitive to lift-off variations and wobble than the more conventional eddy current probes.

A defect lying in the same plane as induced eddy currents has little or no effect on the impedance of the coil because there is virtually no diversion of the electromagnetic field. Consequently, shallow laminar defects parallel to the surface of a metal object cannot be detected with an ordinary surface-scanning coil when its axis is perpendicular to the surface. However, defects of this nature can be detected by a gap probe; an example consists of a ferrite yoke through which a magnetic flux is excited by a pair of encircling coils (Figure 5.5). When the lower end of the yoke is in contact with the metal surface, the metal surface completes the magnetic circuit and eddy currents appear, flowing in

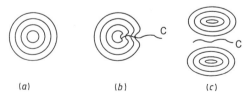

**Figure 5.4** Lines of eddy current flow in a sample tested by self-nulling eddy current probes: (a) defect C absent; (b) defect C partly penetrating; (c) defect C fully penetrating. (Reprinted from Wincheski *et al.*, 1994, by permission of the American Society for Nondestructive Testing).

# Probe design

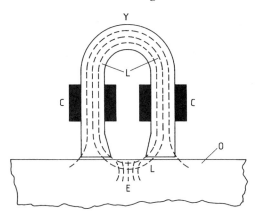

**Figure 5.5** Gap probe: C = coil, Y = ferrite yoke, O = object under test, L = lines of magnetic flux, E = planes of eddy current flow. (Reprinted by permission of Atomic Energy of Canada Ltd)

planes perpendicular to that surface. This probe has the added advantage of close electromagnetic coupling with the object being tested.

Clark and Bond (1989) have investigated the possibility of using a short horizontal coil with its axis parallel to the surface for detecting surface-breaking cracks. The lines of flux follow curved paths and, except in the region below the centre of the coil, the eddy currents travel in planes inclined to the surface and can thus be resolved to provide a component parallel to the surface. But this type of coil produces a lower degree of resolution for flaw detection because of the greater area presented to the surface of the test object. However, it is said to be less sensitive to variations in lift-off compared with the conventional type of surface-scanning probe. It does appear, however that a horizontal probe might well be useful for detecting subsurface laminar defects (cf. the gap probe).

Mention should be made of the saturation probe used for the eddy current testing of ferromagnetic metal cylindrical tubes and bars under conditions of magnetic saturation. It consists of two encircling coils, the outer carrying a DC saturation current and the inner acting as the eddy current detector. Difficulties arise when using a comparable device for surface scanning because the flux lines spread out rapidly at the base of the magnetizing coil. However, where sufficient room is available, an effective method of obtaining saturation is to apply the necessary field with either a strong permanent magnet or a DC yoke to provide a closed magnetic circuit (cf. section 3.3.3).

These subject of eddy current probes is discussed further in section 5.4.3.

## 5.4 REQUIREMENTS FOR EDDY CURRENT MEASUREMENTS

### 5.4.1 General considerations

Basic eddy current applications fall into four principal categories:

- measuring electrical conductivity
- determining magnetic permeability
- evaluating dimensions
- defect detection and sizing

The choice of equipment used depends on the nature of the materials to be tested and the required degree of sensitivity of the measurements. Some applications, such as determining the electrical conductivity of a homogeneous non-ferrous metal and the thickness of a non-metallic coating on a metal base, require only a simple inductance bridge, a phase-sensitive detector, a simple coil and a suitable calibration block. Others, such as crack detection in intricate welded components, often need more sophisticated equipment including specially designed coils and a highly sensitive multifrequency eddy current instrument, perhaps incorporating a minicomputer. The minicomputer can process data at the time of measurement with reference to existing data provided by previous calibrations, made either experimentally or from a computer program using a finite element numerical method. The more basic applications are covered in sections 5.5 and 5.6.

### 5.4.2 Standards for eddy current testing methods

Present standards for eddy current testing procedures are concerned principally with tube testing and measurements of electrical conductivities and coating thicknesses. British (BS) and American (ASTM) standards for eddy current testing are listed in Appendix B.

### 5.4.3 Probe arrangements

The American Society for Testing and Materials (ASTM E 566) defines two basic types of probe arrangement, i.e. absolute coil and comparator coil. With the absolute coil arrangement, the probe contains a single detector coil and, for surface scanning, it is initially applied with a given degree of lift-off to a standard sample having known properties; the output of the instrument is then observed. Following this, the probe is applied with the same degree of lift-off to the object to be tested and the indications of the instrument compared with those obtained for the standard. With the encircling and internal axial coil techniques, the fill-factors for the standard and test samples should

## Requirements for eddy current measurements

both be the same. The author suggests that, where possible, the instrument should be balanced when the probe is applied to the standard sample and any deviations from balance should be observed when it is applied to the test samples, thus improving the sensitivity of detection.

Two coils are used with the comparator coil method, i.e. the reference coil, where the standard sample is retained for the duration of the test, and the test coil, applied to the objects under test. The coils nominally have the same impedance and form arms 3 and 4 of the inductance bridge (Figure 2.12b). Balance is best obtained initially with both coils at the standard sample. The out-of-balance indications of the detector thus indicate how the properties of the test samples differ from those of the standard.

A development of the comparator coil is the autocomparator coil, usually known as the differential coil (or probe) (Figure 5.6), which is used in the scanning mode. The pair of coils, either encircling (Figure 5.6a), internal axial or surface scanning (Figure 5.6b), are located adjacent to one another and are thus applied to two neighbouring parts of the same object. The coils $L_3$ and $L_4$ (Figure 5.6d) are connected in series, with a common terminal (A) between them and are wound in antiphase with respect to one another. With the inductance bridge method, the adjacent coils are located, respectively, in arms 3 and 4 (section 5.2.1). Balance of the bridge is obtained when the properties of the test object are identical at the locations of both coils. A variation of one of these properties, e.g. a surface crack, affects each coil in turn as the probe moves across it,

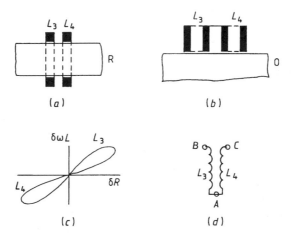

**Figure 5.6** Differential probe: (a) encircling, (b) surface scanning, (c) typical impedance vector pattern for a surface defect, (d) typical circuit diagram. $L_3$ and $L_4$ are coils; A, B and C are coil terminals; R is a cylindrical object under test and O is a flat-surfaced object under test.

and because the windings are in antiphase, the indications of the impedance changes of $L_3$ and $L_4$ in turn take the form shown in Figure 5.6c, i.e the vertical height of the observed signal is double the height obtainable with a single coil passing the same current at the same frequency. Here are some of the advantages of using a differential probe:

1. There is a doubling of sensitivity over that obtained with a single-coil probe, in view of the simultaneous appearance of output signals having equal and opposite phase.
2. The output signal arising from an unwanted gradual variation of say lift-off is eliminated, provided the coils are sufficiently close to one another, e.g. contained in a single housing; the change in impedance is the same for both coils, but these changes cancel out because the coils are wound in antiphase to one another.
3. Impedance variations caused by temperature changes are eliminated because both coils in the probe should be at the same temperature.

The differential coil is intended to be used for the detection of abrupt changes in material properties and is clearly not suitable for detecting extended defects parallel to the surface, e.g. laminations and wall thinning due to corrosion in metal tubes.

A highly sensitive device is the reflection probe used originally to measure thicknesses of metal layers (Dodd, 1977) and more recently to evaluate subsurface defects in metals (Tilson and Blitz, 1985). It is a modified surface-scanning transformer probe containing two secondary coils $L_1$ and $L_2$, having equal inductances (Figure 5.7a and b), connected in series with one another and earthed at the point of connection. They are wound in opposition and, when remote from any metal and properly balanced, the resultant output potential difference $V_o$ is zero. On placing the probe on the surface of the object under test, differences occur in the inductances $L_1$ and $L_2$ because of the lower linkage of magnetic flux through $L_2$, which is remote from the object, so $V_o$ has a finite complex value. For low values, the components of the variations in the impedance of the probe circuit are proportional to those of $V_o$, similar to an inductance bridge (section 5.2.3).

The reflection probe can be used with a commercial eddy current instrument containing an inductance bridge, provided it is suitably modified. This can usually be achieved by disconnecting the arms of the bridge and connecting the primary coil $L_0$ to the input terminals and the secondary coils $L_1$ and $L_2$ in series across the detector arm (Tilson and Blitz, 1985). The output of the instrument can be calibrated in a similar manner to the inductance bridge using a small resistor $\delta R$ (section 5.2.3), i.e. equations (5.2) are applicable but with different values of the constant $A$.

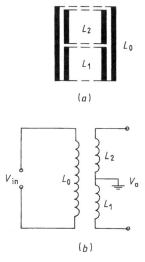

**Figure 5.7** Reflection probe: (a) a vertical section through a diameter and (b) the circuit diagram. $V_{in}$ = input potential difference, $V_o$ = output potential difference, $L_0$ = primary coil, $L_1$, $L_2$ = secondary coils. (Reprinted from Dodd, 1977, by permission of Academic Press)

The reflection probe possesses a further advantage because changes in phase of the coil impedance remain virtually constant with changes in lift-off.

### 5.4.4 Choice of frequency

The choice of the excitation frequency depends on the degree of sensitivity necessary to measure the required parameter (e.g. defect size, electrical conductivity) and also the need to eliminate unwanted changes in impedance caused by factors such as uncontrollable probe lift-off and the presence of edges. Although the optimum frequency can be selected after trial-and-error tests have been made with a standard test block at different frequencies, much time can be saved by studying the impedance data obtained by analytical and modelling techniques (Chapter 4). Consider the simple application of measuring the electrical conductivity of a non-ferromagnetic metal cylinder using an encircling coil. It is clear from Figure 4.2 that a higher change in impedance of the coil for a given change in conductivity $\sigma$ is found to occur at lower frequencies. However, the Figure 4.2 shows that, for very low values of normalized frequency $f_0 = \omega\mu\sigma b^2$, the phase angles between the impedance vectors for $\delta f_0$ (or $\delta\sigma$ at constant $\mu$ and frequency) and $\delta D$ (i.e. change in fill-factor) are so small that phasing out the fill-factor leaves only a small component of impedance to characterize electrical conductivity. On the other hand, when $f_0$ is about 15, the phase angle between the two vectors is sufficiently large to obtain acceptably high sensitivities for conductivity measurements after the fill-factor vector has been phased out. With surface-scanning probes, the resolution between the conductivity and lift-off vectors is generally better than with encircling probes, and the choice of the optimum frequency presents less of a problem (Figures 4.9 and 4.10).

Figures 4.14 and 4.15 indicate that, with either surface-scanning or encircling probes, there is a very small phase angle between the impedance vectors for changes in the size of surface cracks and in the conductivity, hence resolution becomes difficult. However, this problem can be overcome by observing the characteristics of the output signal from the detector. When variations of impedance during scanning are caused by the presence of cracks they are more abrupt than variations caused by changes in conductivity, usually a bulk phenomenon, but they are not necessarily more abrupt than those caused by changes in magnetic permeability, perhaps produced by stress or hardening. These changes may be highly localized and difficulties may be encountered when testing objects made from ferromagnetic materials. Notwithstanding these observations, the detection of subsurface phenomena, e.g. shallow non-surface-breaking defects, requires low frequencies to overcome the restrictions imposed by the skin effect, as characterized by the standard penetration depth $\delta$ (equation 2.90). Note that equation (2.90) applies only to the propagation of plane waves, so it does not take into account any divergence of flux, effectively reducing the amount of penetration. The depth at which the effects of eddy currents can be detected also depends on the sensitivity of the testing equipment and the radius of the exciting coil. A typical value of this depth is $3\delta$ but it may well exceed $5\delta$ with a highly sensitive instrument. Comparing equation (2.90) with the expression for the normalized frequency of surface-scanning coils, i.e. $f_0 = \omega\mu\sigma r_0^2$, where $r_0$ is the coil radius, we see that

$$f_0 = 2r_0^2/\delta^2 \tag{5.4}$$

i.e. the relationship between the normalized frequency and the standard penetration depth depends only on the radius of the coil, and the use of equation (5.4) should be of great help in selecting the optimum frequency for the required degree of penetration. Note that the inductive component $\omega L$ of the coil increases with the product $\omega r_0^2$ (equation 5.3).

Dimension measurements based on changes in either lift-off or fill-factor are best made at higher frequencies where impedance changes are less sensitive to variations of electrical conductivity and magnetic permeability (Figures 4.2, 4.3, etc.) and where there is an increase in sensitivity in measuring $\omega L$. However, the restrictions indicated by equation (5.4) should be considered when the through-penetration technique is used for determining metal thickness.

## 5.5 BASIC EDDY CURRENT TESTS: MEASUREMENTS

### 5.5.1 Electrical conductivity measurements

Measurement of the electrical conductivity $\sigma$ of a non-ferromagnetic metal is fairly straightforward both when obtaining absolute values for homogeneous materials and relative values for objects containing structural variations, e.g. localized heat treatment. The principles of the measurement are based on the variations of impedance with electrical conductivity at a fixed frequency with constant lift-off (or fill-factor) (Figures 4.2, 4.9, etc).

Instruments used solely for conductivity measurements are simple in design and they are mainly used with surface-scanning probes of larger diameter, e.g. 10 mm and more, and typically of the pancake type. They often provide a choice of only a few fixed frequencies in the lower kilohertz range, e.g. 5 and 10 kHz, depending on the required depth of penetration, with a meter serving as the detector output and a phase changer for eliminating lift-off; this is essential for testing metals with rough surfaces. The meter is calibrated (section 5.2.2) using samples having known values of the electrical conductivity $\sigma$ and it can thus be graduated directly in $MS\,m^{-1}$ or per cent IACS.

An alternative method, suitable for an instrument with a cathode-ray tube display, is to observe the lift-off curve obtained from a surface-scanning probe at a given frequency. Figure 4.11 shows how the shapes of the lift-off curves vary with $\sigma$. Suitable lift-off curves, obtained from samples of several metals having known values of $\sigma$, can be drawn on a transparent sheet placed over the oscilloscope screen to serve as a calibrated graticule. The conductivity for the sample under test can then be evaluated from the measured lift-off curve. This method has the obvious advantage that the elimination of the lift-off effect is unnecessary.

Difficulties may arise when attempting to measure the electrical conductivity of a magnetically unsaturated ferromagnetic metal because the expression for $f_0$, where $f_0 = \omega\mu\sigma r^2$, contains the product $\mu\sigma$ which is difficult to resolve (section 5.5.2). The only practical solution to this problem is to apply a saturation magnetic field, where feasible, when the eddy current measurements are made.

Complete eddy current penetration of thin metal sheets may be possible at a sufficiently low frequency, e.g. of up to a few kilohertz (equation 2.90). This phenomenon is applied to the through-transmission method (Förster and Libby, 1986) for which fork probes are used, and the ratio of the complex impedance of the upper coil to that of the lower coil is measured (Figure 5.1d). For non-ferromagnetic metal sheets having constant thicknesses, it is sufficient to measure the relative amplitudes of the potential differences across the coil, after a suitable calibration is made. Complications may arise when the thicknesses of the sheets are no longer constant and reference should be made to Förster's impedance analysis for transmitted eddy currents (Förster and Libby, 1986). The relevant impedance curves bear a superficial resemblance to those for encircling and surface-scanning coils and are characterized by a normalized frequency $f'_0$ given by

$$f'_0 = \omega\mu\sigma t d/2 \qquad (5.5)$$

where $t$ is the plate thickness and $d$ the axial distance between the two coils. By measuring the two impedance components, or the corresponding complex potential difference across the detector, the values of the electrical conductivity and plate thickness can be determined independently of one another, assuming a constant value of magnetic permeability, e.g. $\mu_0$. Alternatively, impedance vectors which characterize variations of plate thickness can be phased out. The principles of this technique can be derived from the analysis by Dodd and Deeds (1968) for two parallel conducting plates, one on top of the other, with air replacing the lower conductor, as indicated in the latter part of section 4.5.2.

The through-transmission method is highly suitable for the rapid scanning of large areas of non-ferromagnetic metal sheets made, for example, from aluminium and stainless steel. Especially advantageous for testing on a production line, it also has the advantage that conductivity and thickness measurements can be performed simultaneously.

The value of the electrical conductivity of a metal depends on several factors, including its chemical composition, the nature of its crystalline structure, its mechanical properties and temperature, as well as its electrical properties. When using eddy currents to measure conductivity it is important, for the sake of correctness and accuracy, to ensure that these factors are kept under control, along with other factors such as the geometry, the magnetic permeability, the temperature of the specimen and the temperature and lift-off of the probe.

Suhr and Guettinger (1993) have pointed out how the specimen should have a specified minimum thickness depending on the eddy current amplitude, which varies in the same way as the amplitude of the magnetic field, as indicated by equation (2.86). Equation (2.90) defines the standard penetration depth, and for a non-ferromagnetic metal with magnetic permeability equal to $4\pi \times 10^{-7}$, the value of $\delta$ can be written approximately as

$$\delta = 503/(\sigma f)^{1/2}$$

where $\sigma$ is the electrical conductivity.

However, in practice, the measurable penetration depth $\delta_{\text{eff}}$ is normally taken to be three times this value, i.e.

$$\delta_{\text{eff}} = 1509/(\sigma f)^{1/2}$$

With, for example, the testing of an aluminium alloy sheet 1.4 mm thick and having an electrical conductivity $\sigma = 20\,\text{MS}\,\text{m}^{-1}$, the maximum frequency required for complete penetration is equal to about 60 kHz.

Suhr and Guettinger used a shielded probe (section 5.3), which maximized the output and minimized any edge effects caused by abrupt changes in sample geometry. They were also able to conduct measurements of conductivity in metals having convex surfaces curved in only one direction, e.g. cylinders. Several test pieces were used for each of the chosen metals, and each test piece had a different curvature. The problem caused by lift-off due to curvature was overcome by taking measurements on each of the test pieces and relating them to those made on a plane surface of the appropriate metal. In this way, one could obtain a correction factor $\delta\sigma/\sigma$, found to be about 25% for a 5 mm radius of curvature, decreasing to approximately 10% for a 10 mm radius and approximately 2% for a 40 mm radius, etc., for frequencies in the range 60–480 kHz and electrical conductivities in the range 4–60 MS m$^{-1}$.

Suhr and Guettinger also considered the errors introduced by increases of temperature experienced with a hand-held probe as a result of heat flow from the body of the operator. This would necessitate recalibration.

### 5.5.2 Magnetic permeability measurements

Magnetic permeabilities are normally measured by using purely magnetic methods but the eddy current technique can be more convenient for sorting metals in accordance with material properties that depend on permeability (section 3.1). Section 2.7.1 mentions how the magnetic permeability of a material is not an easy quantity to define because of its dependence on the value of any applied magnetic field and on its previous history. With eddy current testing, the magnetic permeability of the material in the vicinity of the coil is affected by the strength of the exciting current, which may create difficulties when determining the initial permeability $\mu_i$ of unmagnetized samples. Provided the current through the coil is very small, the resulting recoil permeability $\mu_{\text{rec}}$ (section 2.7.1) should not differ greatly from this. The problem becomes less serious for testing magnetized metal samples, provided the amplitude of the magnetic field generated by the exciting probe is very much less than the value of the coercivity of the metal. The magnetic permeability of any calibration sample should be verified using a direct magnetic measurement (Chapter 3).

Tests can be made with any conventional type of probe, although the through-transmission method might be restricted to thin foils because of the greatly reduced penetration depths for ferromagnetic materials. Difficulties which might well be encountered in resolving the effects of magnetic permeability and electrical conductivity make it desirable for the sample to have a conductivity of known value, or at least to be less affected by the mechanical properties of the material than by the permeability.

The eddy current method can be used to measure the magnetic permeabilities of ferromagnetic metal samples in the form of thin wedges by measuring the depth of penetration (Blitz *et al.* 1986). For this purpose, the measured penetration depth $d$, as opposed to the standard penetration depth $\delta$ (equation 2.90), is defined here as

$$d = k/(\mu_r \sigma)^{1/2} \tag{5.6a}$$

where $k$ is constant for a given frequency and probe, thus allowing for divergence of the flux. The standard was an annealed 10° wedge made from an aluminium alloy having an electrical conductivity $\sigma_2$ of 30.3 MS m$^{-1}$. The test sample was an annealed unmagnetized mild steel 2.5° wedge of conductivity $\sigma_1$ equal to 6.8 MS m$^{-1}$. The standard and the sample were scanned from the thin edge, in turn, with the same surface probe excited at the same input level and frequency. The output signals were observed during scanning. The measured penetration depths $d_1$ and $d_2$ were obtained for those positions of the probe where the output signals reached constant levels. Putting $\mu_r = \mu_{\text{rec}}$ in equation (5.6a) and eliminating $k$, we have

$$\mu_{\text{rec}} = \sigma_2 d_2^2 / \sigma_1 d_1^2 \tag{5.6b}$$

where $\mu_{\text{rec}}$ is the relative recoil permeability of the steel. The experimental error was estimated to be $\pm 3\%$ and relative recoil permeabilities of the order of 100 were obtained by this method.

### 5.5.3 Measurements of dimensions

The following dimensions may be measured by eddy currents:

- Cross-sectional dimensions of cylindrical tubes and rods.

- Thicknesses of thin metal plates and foils and of metallic coatings on metallic and non-metallic substrates.
- Thicknesses of non-metallic coatings on metallic substrates.

Let us consider each of them in turn. Dimensions of cylindrical tubes and rods can be measured with either encircling probes or internal axial coils, whichever is appropriate. Figure 4.2 indicates that the relationship between change in impedance and change in diameter is fairly constant at all but very low frequencies. However, the advantages of operating at a higher normalized frequency $f_0$ are twofold in that, firstly, the contribution of any conductivity change to the impedance of the coil becomes less important and, in any case, it can easily be phased out. Secondly, there is an increase in measurement sensitivity resulting from the higher value of the inductive component $\omega L$ of the impedance. Because of the large phase difference between the impedance vectors corresponding to changes in fill-factor and conductivity (and defect size), simultaneous testing for dimensions, conductivity and defects (section 5.6) can be carried out.

Typical applications include measuring eccentricities of the diameters of tubes and rods and the thicknesses of tube walls. Long tubes are often tested by passing them at a constant speed through encircling coils, generally differential, and providing a close fit to achieve as high a fill-factor as possible.

An important application of tube-wall thickness measurement is the detection and assessment of corrosion, both external and internal. Internal probes must be used when the external surface is not accessible, e.g. when testing pipes that are buried or supported by brackets. Success has been achieved in measuring thickness variations in ferromagnetic metal pipes with the remote field technique (section 6.3).

The thickness of a metal plate or foil on a non-metallic substrate can be measured either with the through-transmission method using a fork probe (section 5.5.1) or with a surface-scanning reflection probe (section 5.4.3). Unfortunately, the effects of electrical conductivity $\sigma$ cannot be phased out and it is important to verify that any variations of $\sigma$ over the region of interest are at a sufficiently low level. At the lower values of normalized frequency, where there is detectable penetration, the variations of probe impedance are more sensitive to changes in electrical conductivity. Calibration is made with samples having known thicknesses.

It is possible to measure the thickness of a thin layer of metal on a metallic substrate where there is complete eddy current penetration of the layer but not of the substrate itself, provided the two metals have widely differing electrical conductivities, e.g. silver on lead where $\sigma = 67$ and $10\,\text{MS}\,\text{m}^{-1}$, respectively. Depending on the required degree of penetration, measurements can be made using a single-coil probe or a transformer probe, preferably reflection type. It is usual to calibrate with samples having known electrical conductivities, although Dodd and Deeds (1968) have performed an impedance analysis for two parallel-sided metal plates having their surfaces in contact with one another. The method has also been used successfully for measuring thicknesses of very thin protective coatings of ferromagnetic metals, e.g. chromium and nickel, on non-ferromagnetic metal bases.

Thicknesses of non-metallic coatings on metal substrates can be determined simply from the effect of lift-off on impedance. Contributions to impedance changes due to conductivity variations should be phased out, unless it is known that conductivity variations are negligible, as normally found at higher frequencies (Figure 4.2, etc). This method has widespread use for measuring thicknesses of paint and plastic coatings. Thicknesses of between 0.5 and 25 $\mu$m can be measured to an accuracy of between 10% for lower values and 4% for higher values. The reader is referred to standard ASTM B 244, which draws attention to edge effects, curvature, surface roughness, the need for constant pressure on the probe and also the requirement to note the presence of any foreign particles. The problem of edge effects is discussed in Chapter 6.

**Figure 5.8** (a) Elcometer 300 eddy current coating thickness gauge (minus probe) incorporating a microprocessor.

*Basic eddy current tests: measurements*

Modern eddy current coating thickness detectors are often pocket-sized with the probe resembling a small pencil. They are usually operated by a small battery and provide a digital read-out in the appropriate units. Calibration adjustments, some of which are laid down by standards, e.g. BS EN 2360 (1995) and ASTM B 244 and E 376, may be assisted by the use of an inbuilt microprocessor (Figure 5.8a). Figure 5.8b illustrates the use of this type of detector for monitoring paint thickness in a critical area of the body of a motor car.

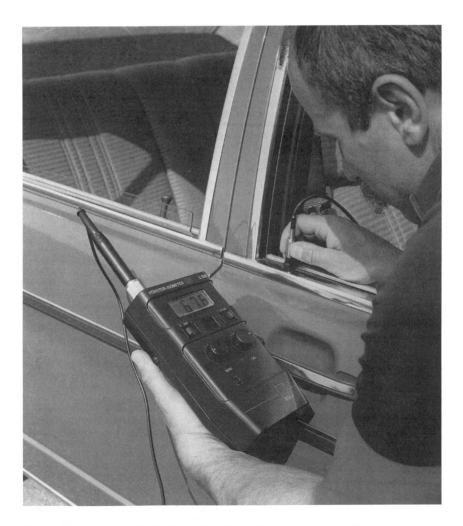

**Figure 5.8** (*continued*) (b) Using the Förster Isometer coating thickness gauge to measure paint thickness on a motor-car body. (Part (a) reprinted by permission of Elcometer Instruments Ltd; part (b) reprinted by permission of the Institut Dr Förster)

156  *Eddy current methods*

## 5.6 BASIC EDDY CURRENT DEFECT DETECTION AND SIZING

### 5.6.1 General considerations

An application of eddy current testing which has rapidly grown in importance is the detection and sizing of surface and subsurface (i.e. shallow) defects in metals. With recent improvements in the design of electrical equipment and probes, the method is both reliable and accurate. Originally eddy current flaw detection was restricted mainly to testing metal cylindrical tubes and rods but it has been rapidly extended to testing objects of almost any shape and size. This has been helped by the development of more recent techniques such as multifrequency testing. Among the more outstanding applications are the testing of turbine blades and bolt holes (e.g. for engine mounting) in aircraft, nuclear reactor cooling tubes and underwater structures and pipelines. The eddy current method is now extensively used as a means of sizing surface cracks in welds. This section deals with the more conventional applications and later developments are covered in Chapter 6. The modelling of cracks is discussed in section 4.6. It is repeated here that, for obvious reasons, planar defects such as cracks and laminations which lie in the planes containing the eddy currents are difficult to find with conventional probes. Defects of this kind, when parallel to the surface, can be detected with the use of a gap probe (section 5.3).

The simplest method of flaw detection is to scan the metal under test with a single-coil probe of appropriate design with a constant degree of either lift-off or fill-factor, and to look for abrupt changes in the output signals resulting from sharp discontinuities in structure, i.e. in the parameter $\mu\sigma$, where $\mu$ is the magnetic permeability and $\sigma$ the electrical conductivity. Figure 4.12 illustrates the diversion of eddy current paths in the presence of a surface crack and Figures 4.14 and 4.15 show the anticipated corresponding changes in impedance. The presence of a small surface crack produces a greater change in amplitude than in phase. However, for larger surface cracks, it has been observed that an increase in size produces a more noticeable change in phase than in amplitude. Note that the maximum crack depth measurable using eddy currents is limited by the penetration depth, for example, when the tip of the crack lies well below this depth.

Recent advances in electronics and instrument technology have allowed small, lightweight, hand-held eddy current flaw detectors to perform much better than earlier generations of full-size equipment operated from the mains. Figure 5.9 is an example of such a device having a frequency range of 100 Hz to 6 MHz and weighing only about 1.25 kg without batteries. The batteries can be carried in a belt worn by

# Basic eddy current defect detection and sizing

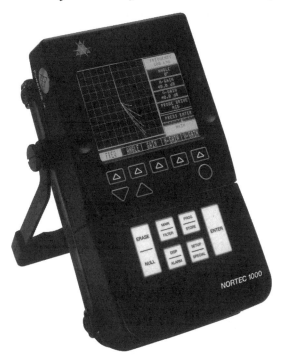

**Figure 5.9** Hand-held eddy current flaw detector. (Reprinted by permission of Staveley NDT Technologies)

the operator. The output is supplied by an electroluminescent display or a liquid crystal display of impedance curves.

### 5.6.2 Defect detection in cylindrical rods and tubes

Eddy current flaw detection of smaller diameter cylindrical metal rods and tubes with encircling or, where appropriate, internal axial probes is widespread and its principles have been extensively studied by Förster and others (section 4.6). The defects most commonly occurring include surface and subsurface cracks and internal defects such as blowholes and inclusions, and for tubes, cracks and corrosion at both internal and external surfaces. In addition, the occurrence of dents in thin-walled tubes creates variations in fill-factor and, where appropriate, changes in wall thickness (section 5.5.3).

The method of calibration is often laid down by national standards (Appendix B) and the reference sample takes the form of a short defect-free length having the same cross-sectional dimensions and properties as the tube or rod being tested. Artificial defects are cut or drilled in the sample. For example, ASTM standard E 268,

Electromagnetic (eddy-current) testing of seamless copper and copper-alloy tubes, relates the procedures applicable to tubes having outside diameters ranging from 6.35 to 50.8 mm and wall thicknesses from 0.889 to 3.04 mm using an encircling coil. This standard is fairly flexible and it allows the following options for artificial defects:

- A round-bottomed transverse notch on the outside of the tube in each of three successive transverse planes.
- A hole drilled radially through the tube wall in each of three successive transverse planes at 0°, 120° and 240°.
- A round transverse notch on the outside of the tube at 0° and another at 180°, and one hole drilled radially through the wall at 90° and another at 270°. Only one notch or hole should be made in each transverse plane.
- Four round-bottomed transverse notches made on the outside of the tube, all on the same element of the tube.
- Four holes drilled radially through the tube wall, all in the same element of the tube.

The various dimensions of these artificial defects, appropriate to the values of outside diameter and wall thickness, are given – but there is no attempt to relate these dimensions to the dimensions of a real defect. They are purely provided as acceptance standards in the same way as flat-bottomed holes with ultrasonic testing (e.g. Krautkrämer, 1983), i.e. a tube should be rejected if it produces a signal of magnitude greater than produced by an artificial defect of a specified size.

The design of either an encircling or internal axial probe depends on the values of the cross-sectional dimensions of the sample and the degree of penetration required, at the appropriate frequency. The appropriate frequency may determine whether a single probe or a transformer probe is selected. Ferromagnetic metals can be tested with a saturation probe, which normally presents few problems with encircling coils, because of the close coupling of the magnetizing coil with the test object.

Larger-diameter rods and tubes, i.e. greater than about 50 mm, are tested more effectively with surface-scanning probes, either singly or in arrays, with the scanning performed in a helical manner.

When either encircling or internal axial coils are used for testing non-ferromagnetic metal tubes, external and internal surface defects can be resolved from one another by considering the respective values of the impedance phase change caused by their presence. This phenomenon has been investigated by Förster (1986) and Figure 5.10 shows the relationships between the ratio $J_i/J_o$ of the current density amplitudes (cf. equation 4.35a) at the inner and outer surfaces, respectively, and the normalized frequency $f_0$ for different ratios

## Basic eddy current defect detection and sizing

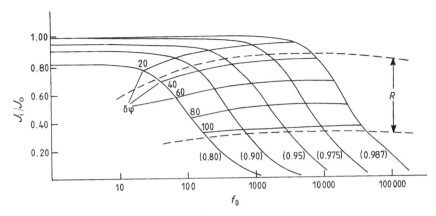

**Figure 5.10** Relationship between eddy current density ratio $J_i/J_o$ and normalized frequency $f_0$ for different values of diameter ratio $d_i/d_o$ (in parentheses) for the testing of a non-ferromagnetic metal tube with an encircling coil, showing differences $\delta\phi$ of impedance phase angles for defects at inner and outer surfaces. The subscripts i and o denote inner and outer surfaces, respectively, and $R$ indicates the range of good resolution. Here $f_0 = \omega\mu\sigma d_i t$ where $t$ is the thickness of the tube. (Reprinted from Förster, 1986, by permission of the American Society for Nondestructive Testing)

$d_i/d_o$ of the internal to external diameters of the tube, ranging from 0.80 to 0.987. Förster defined $f_0$ in terms of $d_i$ and the thickness $t$ of the tube which, when expressed in the form compatible with SI units, is equal to $\omega\mu_0\sigma d_i t$. Figure 5.10 also shows variations of $J_i/J_o$ with $f_0$ for different values of the phase difference $\delta\phi$ between $J_i$ and $J_o$, which indicates that good resolution for measuring the respective impedances of the coil occurs when $\delta\phi$ lies between 40° and 100°. Optimum sensitivity arises with $J_i/J_o$ between 0.4 and 0.5. Thus, when $d_i/d_o = 0.8$, the optimum value of $f_0$ is in the region of 100, but for a very thin tube, i.e. when $d_i/d_o = 0.987$, $f_0$ reaches a value of the order of 20 000, very much greater than the optimum value of 15 for cylindrical rods (section 4.6.2).

When $\delta\phi$ is less than 40° the resolutions between indications originating from internal and external defects become too low for practical use, and when $\delta\phi$ exceeds 100° the detection of internal defects becomes difficult because of low values of $J_i/J_o$, hence low amplitude resolution.

The basic eddy current method is not usually suitable for testing unsaturated ferromagnetic tubes for defects, except perhaps very thin-walled tubes, because of the decrease in penetration as a result of the high values of relative permeability $\mu_r$. When magnetic saturation is impractical e.g. with larger and thick-walled tubes, recourse should be made to a magnetic flux leakage technique (Chapter 3). Rapid advances are being made in the development of the remote field eddy current method for testing ferromagnetic tubes with internal axial coils, but several difficulties have still to be overcome (section 6.3).

### 5.6.3 Surface scanning for defects

The use of surface-scanning probes is required for the precise location of discontinuities and they are most effective with plane surfaces. With curved surfaces it is important to make the diameter of the probe small enough to ensure the surface beneath it is effectively plane, so that lift-off is constant to within the accuracy required for the measurement, e.g. to within 0.1 mm for measuring crack sizes.

Calibration is usually performed with the aid of a test block made from the same material as that being tested and containing vertical saw-cuts of different depths. It has been shown (Blitz and Alagoa, 1985) that, when a single-coil probe scans a surface containing a crack, the maximum value of impedance attained by the coil is determined only by the vertical depth of the crack tip below the surface; it is independent of the width of the crack as defined in section 4.6.2 (see also Figure 4.13). Figure 5.11 illustrates variations of the impedance of a single-coil probe during a scan, at a constant degree of lift-off, over the surface of a metal block containing simulated cracks having equal depths but different angles of inclination. The impedance locus comprises a curved line for a perpendicular 'crack' (Figure 5.11a) but opens out into a loop when the 'crack' is inclined to the surface; the area of the loop increases with the angle of inclination (Figure 5.11b and c). The shape of the loop is determined by the shape of the 'crack' and Figure 5.11d shows how a loop can take the form of a figure of eight when the simulated crack is bent.

Note that a simulated crack such as a saw-cut may not always serve as a realistic model of a real crack, even if the real crack has plane and parallel sides, because of practical difficulties in attaining the small thicknesses which are characteristic of most cracks. In practice, a saw-cut may be up to 10 times thicker than the crack it is intended to simulate, although improvements in this respect can be achieved with the use of an electron beam cutter. The disturbance of the eddy current pattern resulting from a real crack may thus be considerably narrower than the disturbance from a simulated crack having the same depth; when it is scanned, the simulated crack provides a narrow impedance loop rather than a line. However, it is often convenient to characterize a real defect with reference to a standard saw-cut, i.e. an acceptance standard.

The maintenance of a constant degree of lift-off may be difficult in practice because of **probe wobble** with manual scanning and also as a result of uneven surfaces with both manual and automatic scanning.

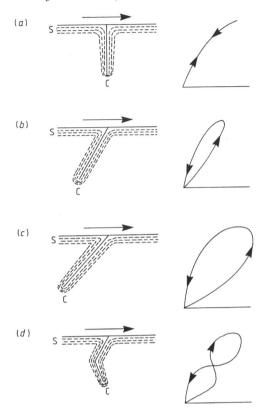

**Figure 5.11** Examples of cathode-ray oscilloscope displays with an eddy current flaw detector for variation of the components of the impedance of a single surface probe scanning a metal test block containing saw-cuts to simulate cracks, each of similar depth below the surface (S), with the phase adjusted to provide a horizontal line for the lift-off impedance vector (L/O). Also shown are diagrams of the corresponding saw-cuts and vertical sections of eddy current flow: (a) perpendicular saw-cut, (b) saw-cut inclined at a small angle, (c) saw-cut inclined at a larger angle and (d) saw-cut with a bend. Arrows indicate the directions of scan.

Errors occurring as a result of lift-off instability can be reduced by using differential probes (section 5.4.3) but in this context the value of the change in impedance due to the presence of a defect, i.e. the defect signal, depends on the amount of probe lift-off. Small lift-off changes, such as those caused by uneven surfaces and probe wobble, do not affect the defect signal to any significant extent when differential coils are used. Furthermore, variations of electrical conductivity and magnetic permeability are not likely to produce any resultant signal because changes in these quantities are gradual and, except for large-diameter coils, e.g. pancake coils, variations in impedance due to these factors affect both coils virtually simultaneously.

CHAPTER 6

# More advanced eddy current testing methods

## 6.1 AUTOMATIC TESTING

As with flux leakage (section 3.6) and many other NDT methods, the eddy current technique lends itself readily to automatic testing, often involving the use of a computer to locate and monitor the positions of the probes and to process the data collected by them; both the speed and reliability of the testing are thereby increased. A simple example is the high-speed rotating probe used for testing defects in tubes. The probe coil is located at the end of a diametrical beam or on a cylinder which rotates by electrical propulsion around the tube axis and follows a helical path as it progresses through the tube in a corkscrew manner. This device can also be used to test for surface defects in fastener holes such as those located in aircraft bodies (Figure 6.1).

**Figure 6.1** Using a rotating probe to test defects in fastener holes of an aircraft wing. (Reprinted by permission of the Institut Dr Förster)

## Multifrequency testing

A more sophisticated arrangement is an encapsulated waterproof probe assembly (Figure 6.2) devised for the internal scanning of heat exchanger tubes as used for chemical plants, electricity generators and nuclear energy installations (Neumaier, 1983). The particular assembly illustrated has a diameter of 19 mm, to fit the tube as closely as possible, and a length of 190 mm. The cable at one end is connected to the power supply and instrumentation. The power of the motor which drives a disc housing a differential probe arrangement is kept as low as possible so as to minimize vibrations, hence minimizing possible friction between the probe and the surface of the tube wall. The provision of a rotating transformer eliminates the need for sliding contacts with the terminals of the probe coils.

### 6.2 MULTIFREQUENCY TESTING

The impedance of an eddy current probe may be affected by several factors:

- Variations in operating frequency.
- Variations in the electrical conductivity and the magnetic permeability of the test object as caused by structural changes, e.g. grain structure, work hardening, heat treatment, etc.
- Changes in lift-off or fill-factor resulting from probe wobble, uneven surfaces and eccentricity of tubes caused either by faulty manufacture or denting.
- The presence of surface defects, such as cracks, and subsurface defects such as voids and non-metallic inclusions.
- Dimensional changes, e.g. thinning of tube walls due to corrosion, deposition of metal deposits or sludge, and the effects of denting.
- The presence of supports for tubes, e.g. brackets.
- The presence of discontinuities such as edges.

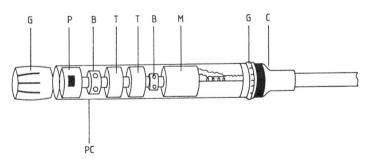

**Figure 6.2** Encapsulated eddy current probe assembly for the internal scanning of metal tubes: C = plug with cable in protective sleeving, G = guide, M = motor, B = bearing, T = rotating transformer, P = probe disc, PC = protective casing. (Reprinted by permission of the British Institute of Non-Destructive Testing)

Two or more of these factors are generally present simultaneously. In the simple case where interest is confined to detecting defects or other abrupt changes in geometry, a differential probe can be used to eliminate unwanted factors, provided they vary in a gradual manner, e.g. variations in electrical conductivity and tube thinning, and therefore affect both coils of the probe simultaneously. If, however, it is necessary to characterize these gradual variations, a single-coil probe should be used.

Two independent parameters can usually be measured at a given frequency by phasing out the corresponding impedance components one at a time. If a larger number is required, additional measurements are necessary at one or more different frequencies, the values of which depend on their nature, as discussed below. Ideally it should be possible to determine the values of $2n$ parameters by operating at $n$ well-spaced frequencies, but this is only possible if the phases of the corresponding vectors can be adequately resolved.

Figure 6.3 illustrates a calibration tube, with a support plate, which simulates a steam generator tube. It is scanned by an internal axial probe and contains simulated defects in the forms of inside and outside grooves, a through-hole, a dent and a magnetic deposit together with the corresponding impedance vectors of both absolute and differential probes. The operating frequency is 250 kHz which, for this particular tube, is said to be equal to $f_{90}$, i.e. the frequency at which the vectors corresponding to depth variations of the inside and outside grooves are initially perpendicular to one another. The impedance vector corresponding to a decreasing fill-factor happens to lie in the same direction as the changes in depth of the outside groove. The impedance vector variations illustrated in Figure 6.3b and c are independent of one another; each corresponds to a single factor when all the others are absent. With the exception of the rare case when there is a phase difference of **exactly 90°** between two impedance component vectors, the observed pattern arising from the simultaneous appearance of two or more factors giving rise to impedance changes can only be resolved into its components by suitable frequency, amplitude and phase manipulation. This can be achieved through multifrequency testing.

The principles of the method can be simply explained (Cecco *et al.* 1981) with reference to the testing of the calibration sample of a steam generator (Figure 6.3). Figure 6.4 illustrates the impedance vectors for a single-coil probe excited at three selected frequencies, i.e. $f_1 = 20$, $f_2 = 100$ and $f_3 = 500$ kHz. Only dents, holes and internal defects are detectable at the highest frequency $f_3$, and at the lowest frequency $f_1$ the effects of the baffle plate and the magnetite deposits predominate, although defects are just about detectable. Magnetite deposits are characterized by high values of magnetic permeability.

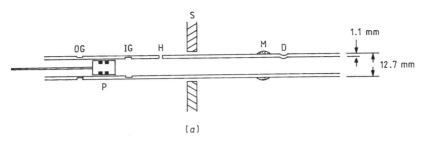

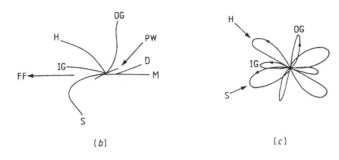

**Figure 6.3** (a) Typical calibration tube having an outside diameter of 12.7 mm, wall thickness of 1.1 mm and electrical conductivity of 1.02 MS m$^{-1}$; it contains artificial defects tested with an internal axial probe (P). (b) Pattern of impedance vectors with a single-coil probe. (c) Pattern of impedance vectors with a differential probe. The test frequency $f_{90} = 250$ kHz. OG = outside groove, IG = inside groove, H = through-hole, S = steel support plate, M = magnetite deposit, PW = probe wobble, FF = direction of decreasing fill-factor, P = probe. (Reprinted from Cecco et al., 1981, by permission of Atomic Energy of Canada Ltd)

To test for a dent in the neighbourhood of a baffle plate and the magnetite deposit, the undesirable signals originating from these items should first be eliminated. Figure 6.5a indicates how this is done by combining the individual signals for the frequencies $f_1$ and $f_2$. The signal for $f_1$ is reduced in size and the signal for $f_2$ is rotated in phase until they appear to be identical to one another in size and shape. By performing a frequency subtraction to give $c_1 = f_2 - f_1$, whereby the phase of the frequency $f_1$ is rotated by 180°, the impedance signal is virtually cancelled out, leaving only residuals due to the elimination not being 100% complete. This may not matter when testing for dents but it could mean that small cracks underneath the baffle plates are undetectable. If required, the dent signal can be suppressed by similarly combining the signal for the frequency $f_3$ and the signal for the mixture $c_1$ then subtracting to give $c_2$, i.e. $c_1 - f_3$ (Figure 6.5b). For this frequency mixture, cracks can be detected and measured, although it has been pointed out by

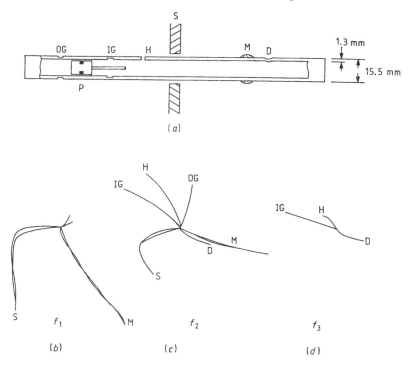

**Figure 6.4** (a) Internal probe scanning calibration tube with single-coil probe; (b) impedance vectors at a scanning frequency $f_1 = 20$ kHz; (c) impedance vectors at $f_2 = 100$ kHz; and (d) impedance vectors at $f_3 = 500$ kHz. The symbols are as for Figure 6.3 and $f_{90} = 130$ kHz. (Reprinted from Cecco et al., 1981, by permission of Atomic Energy of Canada Ltd)

Cecco et al. that multifrequency detection is more effective for outside cracks than for inside cracks. The signals from the inside cracks are affected by what is known as ID noise due to cyclic internal diameter variations and probe wobble.

The multifrequency method can be performed with either a single coil or a differential probe (Davis, 1981). The differential probe is more suitable for detecting small volume defects such as pits and cracks, and the single coil is more suitable for detecting large volume defects and support cracks as well as for sizing dents.

Figure 6.6 illustrates a multifrequency eddy current instrument designed for inspecting heat exchanger tubing. It can generate up to four independently selected frequencies and can display simultaneous x–y signals onto four screens. Satisfactory operation is possible with probe speeds of up to $2\,\mathrm{m\,s^{-1}}$. The instrument can be computer controlled and can be adapted to store data on optical disks.

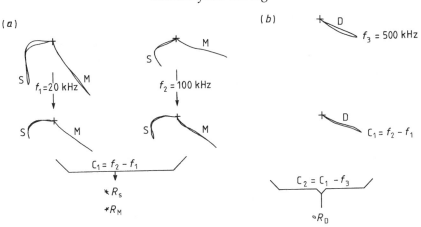

**Figure 6.5** (a) Frequency subtraction for signals from baffle plate (S) and magnetite deposits (M) after vertical axis compression by 0.74 for $f_2$ and phase rotation of 19° for $f_2$. $R_S$ and $R_M$ indicate residual signals for S and M, respectively. (b) Frequency subtraction for suppression of a signal from a dent: $R_D$ = residual dent signal. (Both parts reprinted from Cecco et al., 1981, by permission of Atomic Energy of Canada Ltd)

## 6.3 REMOTE FIELD TESTING

Eddy current testing for external defects in tubes when external access is not possible, e.g. with buried pipelines, is conducted using internal probes. When testing thick-walled ferromagnetic metal pipes with conventional internal probes, very low frequencies (e.g. 30 Hz for a steel pipe 10 mm thick) are necessary to achieve the through-penetration of the eddy currents; this produces a very low sensitivity of detection. The degree of penetration can, in principle, be increased by the application of a saturation magnetic field, but because of the large volume of metal present, a large saturation unit carrying a very heavy direct current may be required to produce an adequate saturating field. However, it is possible to achieve a greater penetration by using pulsed saturation (section 6.5).

The difficulties encountered in the internal testing of ferromagnetic tubes can be greatly alleviated with the use of the remote field eddy current method, which allows measurable through-penetration of the walls at three times the maximum frequency possible with the conventional direct field method. This technique was introduced by Schmidt (1984, 1986, 1989) in 1958, but although it has been used by the petroleum industry for detecting corrosion in their installations since the early 1960s, it has only recently evoked general interest. It is highly sensitive to variations in wall thickness but relatively insensitive to fill-factor changes (sections

168    *More advanced eddy current testing methods*

**Figure 6.6** Computer-controlled ect MAD multifrequency eddy current instrument. (Picture: Gene Woolridge; reprinted by permission of Eddy Current Technology Incorporated)

4.4 and 5.5.3). The method has the added advantage of allowing equal sensitivities of detection at both inner and outer surfaces of a ferromagnetic tube but it cannot differentiate between signals from these respective surfaces.

In its basic form (Figure 6.7), the probe arrangement consists of an exciting coil (A) and a receiver (B) kept at a rigidly fixed separation along the axial direction. Receiver B is depicted as an internal axial coil, but it can be an array of surface-probe coils equally spaced around the circumference of a circle having a radius compatible with the minimum practical degree of lift-off for each of the probes. The separation between A and B should be at least twice the inner diameter of the tube, preferably two and a half times, for the reasons explained below.

Coil A induces a magnetic field in the normal manner; some of the field penetrates the wall of the tube and the rest remains within the tube's air space. Eddy currents following circular paths concentric with the axis of the tube flow within the tube wall and set up a reverse magnetic field; the reverse field attenuates that part of the

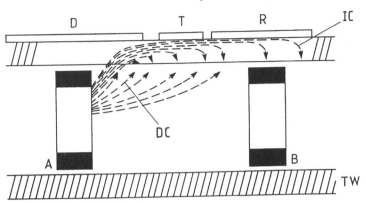

**Figure 6.7** Arrangement of probes inside a ferromagnetic tube for remote field eddy current testing. A = excitation coil, B = receiving coil or array of coils, D = direct field zone, T = transition field zone, R = remote field zone, DC = direct coupling, IC = indirect coupling. Dashed lines indicate the flow of the magnetic field. (Reprinted from Brown and Le, 1989, by permission of the American Society for Nondestructive Testing)

field remaining within the air space, which decreases to zero before reaching receiver B. The region where the field induced directly by the exciting coil A is active is called the direct field zone and this field can produce a current in any coil suitably placed within the zone. The remote field zone is the region in which no direct coupling can take place between A and any coil or array B inside it. Coupling can take place only through diffusion of the magnetic flux excited by A into the tube wall and its subsequent spreading lengthwise along the tube, but with a lower attenuation than for the direct field.

Sullivan *et al.* (1990) have investigated the variations of both amplitude and phase for the magnetic vector potential $A$, as defined by equation (2.25), with axial distance from the exciting coil expressed in multiples of the internal diameter (ID) of the tube (Figure 6.8a and b). The amplitude of $A$ inside the pipe initially decreases rapidly, i.e. within the direct field zone, to a minimum value at a distance of 1.7 IDs, where there is a sharp discontinuity in the value of its phase. From this point there is an abrupt increase of amplitude to a maximum beyond which, at about 2.5 IDs, there is a linear and gradual decrease. The amplitude of $A$ at the outside surface of the pipe decreases steadily in the absence of any discontinuities, as does the phase. It is quite clear the limit of the direct field zone occurs at the distance of 1.7 IDs from the exciting coil. The optimum position of the detector thus lies about 2.5 IDs from the exciting coil.

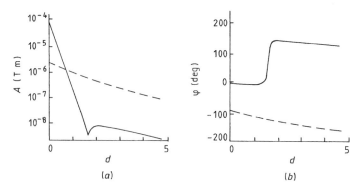

**Figure 6.8** Variations of (a) amplitude $A$ and (b) phase $\phi$ of the magnetic vector potential with axial distance from the excitation coil, as measured in terms of the number of internal diameters of a steel tube having an electrical conductivity of 6.67 MS m$^{-1}$, a relative magnetic permeability of 70, an internal diameter of 76.2 mm and a wall thickness of 7.62 mm when undergoing remote field eddy current testing at a frequency of 40 Hz, for which the standard penetration depth $\delta$ is 3.68 mm. Continuous and broken lines represent variations of $A$ and $\phi$ at inner and outer surfaces, respectively. (Reprinted from Sullivan *et al.*, 1990, by permission of the British Institute of Non-Destructive Testing)

The remote field technique has been highly effective in testing tube-wall thinning but, in its present form, it is not suitable for crack detection. However, Atherton *et al.* (1989) have achieved some success in increasing the flux penetration through the tube wall by using 'saturation windows', whereby permanent magnets are located in the vicinities of the wall at the two probe positions, thus increasing the sensitivity of the method and enabling it to detect cracks.

Dubois *et al.* (1992) report that working in the transition zone can increase sensitivity in measuring defects, allowing the probe length to be shorter and enabling a higher degree of resolution. The resultant field effect becomes a maximum where direct and indirect fields have equal magnitudes and opposite phases. Small variations in the incident magnetic field can produce large changes in the resultant field, thus increasing the sensitivity of defect detection. With a careful choice of frequency it is possible to resolve signals indicating variations of magnetic permeability from signals indicating the presence and size of defects.

The remote field eddy current method has been successfully used to test non-ferromagnetic metals and metal plates where the far surface is inaccessible to shielded probe coils (Sun *et al.* 1996). It is relatively insensitive to probe lift-off and wobble (Mackintosh *et al.* 1996). Frequencies are used in the range 10–200 Hz, depending on the nature of the material being tested.

## 6.4 LIFT-OFF FLAW DETECTION

Variations in lift-off with an eddy current surface-scanning probe often result from the unevenness of the object's surface or from the presence of surface deposits. The effects of small variations in lift-off, such as those caused by probe wobble when testing for flaws, can often be eliminated by a differential probe but this may not be possible with larger variations, as found with very rough surfaces, where the change in impedance caused by the defect is itself a function of lift-off.

It has been shown (section 4.5.2) that the variations of impedance components of an eddy current coil resulting from lift-off changes depend on the value of the normalized frequency $f_0$, where $f_0 = \omega\mu\sigma r_0^2$, for a given frequency $f$ and coil radius $r_0$; they therefore depend on the product $\mu\sigma$ of the magnetic permeability and the electrical conductivity of the test sample. Figure 4.11 shows variations of impedance components of a coil with length 1 mm and radius 1 mm at a frequency of 63.3 kHz for some different values of $\mu\sigma$. Figure 4.14 shows how the size of the phase difference is usually small between the vectors indicating variations of $\mu\sigma$ and defect characteristics, hence it should be expected that a change in the configuration of a lift-off impedance curve may be caused either by a change in $\mu\sigma$ or by the presence of a defect.

Oaten and Blitz (1987) describe eddy current measurements of lift-off from a mild steel test block containing saw-cuts having different depths ranging from 0 to 10 mm. The coil was balanced, in each case, at 'infinite' lift-off from the block and lowered to the surface, firstly at a defect-free region then, in turn, at the opening of each saw-cut. Figure 6.9 shows that, for a frequency of 500 kHz, progressive changes in position of the lift-off curve occur for increase in slot depth $d$ from zero to a maximum value of 10 mm. Thus, to evaluate crack depths in a sample made from a material identical to that of the test block, calibration curves are first plotted for slots of different depths at the desired frequency of operation. If the instrument has a cathode-ray oscilloscope display, these curves can be traced on a transparent sheet placed in contact with the oscilloscope screen, thus providing a graticule which can be readily used to assess depths of surface cracks in the object to be tested (section 5.5.1). Evaluations of crack depths made in this way were compared with evaluations from the same cracks using the AC potential drop (ACPD) method (section 8.2.3) and agreement was found to within 0.1 mm for crack depths of up to 4 mm and to within 0.5 mm for depths of up to 8 mm.

Similar measurements were performed with a test block having the same depths of saw-cuts but made from pipe steel, and a set of lift-off curves similar to those for the mild steel block were obtained,

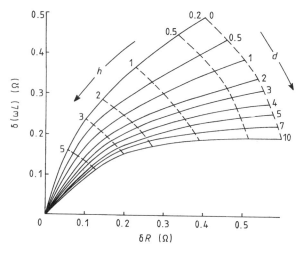

**Figure 6.9** Lift-off curves (full curves) obtained with an unmagnetized mild steel sample with a 13 µH coil excited at a frequency of 500 kHz, with balance obtained for 'infinite' lift-off. Broken curves indicate impedance variations with saw-cut depth $d$ at constant values of lift-off $h$. Values of $d$ and $h$ are in millimetres. (Reprinted from Oaten and Blitz, 1987, by permission of Gordon and Breach)

but each displaced clockwise by an angle $\theta$. The position of the lift-off curve for a defect-free part of the block was determined by the change in value of the parameter $\mu\sigma$. By superposing the two sets of curves and rotating one by the angle $\theta$, coincidence of the curves for the defect-free portions of the samples was obtained. The change of impedance amplitude for a given degree of lift-off, corresponding to a cut of a given depth, from the defect-free value was the same for each metal but the two amplitudes differed in phase by the angle of rotation $\theta$. Figure 6.10a and b show the results of rotation for the curves corresponding to zero defect and 1 mm deep saw-cuts for mild steel and pipe steel. For a lift-off value of 0.2 mm, the respective changes in impedance of magnitudes $AB$ and $AC$ are inclined at an angle $\theta$ to one another.

Using this lift-off or 'touch' method, lift-off curves could be constructed for the saw-cuts in the pipe-steel block from the curves obtained for the mild steel block and for a defect-free sample of pipe steel by performing simple calculations. Defect depths could therefore be assessed with higher confidence in their accuracy than when using more conventional methods, mainly because the uncertainties associated with changes in magnetic permeability were eliminated.

So far, this method has not been successful with non-ferromagnetic metals.

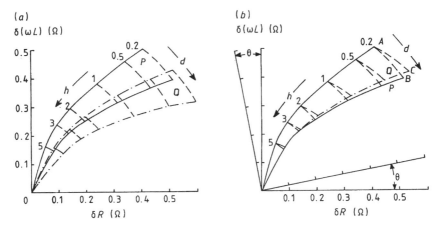

**Figure 6.10** (a) Lift-off curves (full curves) obtained for unmagnetized samples of mild steel (*P*) and pipe steel (*Q*) for defect-free parts of the samples and 1 mm deep saw-cuts. (b) As (a) but with the axes for pipe steel rotated by an angle $\theta$ so the defect-free lift-off curves for both samples coincide. Parameters *d* and *h* are expressed in millimetres. (Reprinted from Oaten and Blitz, 1987, by permission of Gordon and Breach)

## 6.5 PULSED EDDY CURRENT TESTING

The use of pulsed eddy currents has long been considered for testing metals (Libby, 1971) and it has been applied to operations in specialized areas, such as in the nuclear energy industry, where testing equipment is often constructed to order. However, significant progress in this direction has taken place only recently after appropriate advances in technology (Krzwosz *et al.* 1985; Sather, 1981; Waidelich, 1981; Wittig and Thomas 1981) but, at the time of writing, commercial equipment is not yet available. The method has the potential advantages of greater penetration, the ability to locate discontinuities from time-of-flight determinations and a ready means of multifrequency measurement. But, at present, it does not generally have the precision of the conventional methods. The apparatus is somewhat complicated in design and not readily usable by the average operator experienced with the conventional eddy current equipment. Its main successes are in the testing of thin metal tubes and sheets as well as metal cladding, for measuring thicknesses and for the location and sizing of internal defects.

When comparing the pulsed method with the conventional eddy current technique, the conventional technique must be regarded as a continuous wave method for which propagation takes place at a single frequency or, more correctly, over a very narrow frequency bandwidth. With pulse methods, the frequencies are excited over a wide band, the extent of which varies inversely with the pulse length; this allows multifrequency operation. As found with ultrasonic testing, the total amount of energy dissipated within a given period of time is considerably less for pulsed waves than for continuous waves having the same intensity. For example, with pulses containing only one or two wavelengths and generated 1000 times per second, the energy produced is only about 0.002 of that for continuous waves having the same amplitude. Thus considerably higher input voltages can be applied to the exciting coil for pulsed operation than for continuous wave operation.

Pulsed waves can reasonably be expected to allow penetration of measurable currents through a metal sample to a depth of about 10 times the standard penetration depth $\delta$, provided a suitable probe is used, e.g. a shielded ferrite-cored coil (section 5.3). Thus penetration is possible through a 2 mm thick plate at frequencies of 1–3 kHz for non-ferromagnetic metals having corresponding electrical conductivities ranging from 60 down to 20 MS m$^{-1}$. With an unmagnetized steel plate, however, also 2 mm thick, where $\sigma = 5$ MS m$^{-1}$ and $\mu_r = 100$, the maximum frequency for through-penetration is only 100 Hz. Equations (2.89) and (2.90) show that the wavelength of electromagnetic radiation in a metal is equal to $2\pi\delta$, indicating at least half-wavelength resonance in the thickness mode should be possible.

Pulsed eddy currents may be generated by a thyratron connected in series with the exciting coil through a capacitor (e.g. Waidelich, 1981). A direct voltage, of the order of 1200 V, slowly charges the capacitance and, when the thyratron conducts, there is an abrupt discharge through the coil in which free-damped harmonic oscillations occur (section 2.5.2). This is repeated periodically, e.g. at 1 kHz, so as to propagate the eddy current pulses through the metal.

The currents are detected by a receiving probe located either adjacent to or on the opposite side of the metal sample from the exciting probe, when access is possible. The range of propagated frequencies depends on the logarithmic decrement of the exciting circuit, and because the speed of the waves is a function of frequency (equation 2.89), dispersion takes place and the pulse changes in shape as it progresses through the metal. As one would expect, the height of the peak and its time delay can be related to the thickness of the metal. Waidelich reports a maximum penetration of 90 mm for aluminium sheet and 10 mm for steel. For 6 mm thick sheets, the peak value of the received pulse voltage was 13 V for aluminium but only 20 mV for steel. Krzwosz *et al.* (1985) have shown how pulses that result from the presence of internal simulated defects produce broadening with an increase in depth (Figure 6.11).

The frequency content of the pulses depends on their lengths and, in the extreme, contains continuous spectra ranging from less than

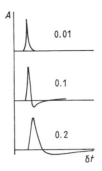

**Figure 6.11** Variation of pulse location and shape with defect depth for pulsed eddy current testing. $A$ is the amplitude and $\delta t$ is the delay time; 0.01, 0.1 and 0.2 are the defect depths in millimetres. (Reprinted from Krzwosz et al., 1985, by permission of Gordon and Breach)

100 Hz to 1 or 2 kHz. By performing a Fourier transformation, the pulse obtained by the receiving probe can be displayed in the form of the variation of amplitude (or phase) with frequency. By sampling different delay times within a pulse, different parts of the spectrum can be evaluated (Sather, 1981). If both amplitude and phase are measured, two parameters (e.g. presence of defects, variations in tube thickness and changes in fill-factor or lift-off) can be evaluated for each frequency selected, in the same way as with the multifrequency method (section 6.2) although, at present, with a lower degree of precision.

Dodd et al. (1988) have designed and developed a pulsed magnetic saturation method for the eddy current testing of ferromagnetic metals. The DC field pulses are generated by passing a high-current pulse through an electromagnet so as to produce saturation in the metal object; the pulse length is made equal to the thickness of the object, thus ensuring complete eddy current penetration where feasible. The DC pulse, of the order of 1 ms duration, simultaneously produces an eddy current pulse which is detected by a probe; the output of the probe is characteristic of the material being tested.

This technique has the advantage of producing high magnetic peak powers with low average powers, thus keeping any heating of the test sample down to an acceptable level. It has been applied successfully for the internal testing of the walls of steel steam generator tubes, and tubes of diameter 10.9 mm and wall thickness 5 mm have been examined with peak powers of 500 kW. Small defects close to the external surfaces can be detected, and by taking advantage of the multifrequency properties of pulsed eddy currents, their indications can be resolved from those that originate from other characteristics of the tubes.

More recent work on the use of pulsed eddy currents has been reported by Gibbs and Campbell (1991), who inspected cracks under fasteners in aluminium aircraft structures. Here a Hall element was used as a receiver. Radial position, approximate depth and relative size of defects, hidden under fastener heads, could be determined in countersunk areas for defect depths of up to 7 mm for non-ferrous fasteners and 14 mm for ferrous fasteners.

Lebrun *et al.* (1975) report the detection of deep cracks in ferromagnetic samples using an emission coil excited by square pulses of high intensity and employing highly sensitive magneto-resistive sensors to measure the resultant magnetic fields. Defects of 1 mm × 1 mm could be detected at a depth of 5 mm and having dimensions of 3 mm × 4 mm at a depth of 20 mm.

## 6.6 MICROWAVE EDDY CURRENT TESTING

The use of electromagnetic microwaves for the non-destructive testing of dielectric materials is fairly well established (Chapter 7) and it can be extended to testing the surface properties of electrical conductors. Values of microwave frequencies lie approximately within the range 300 MHz to 100 GHz, hence standard penetration depths $\delta$ are very small. For most metals they are less than 1 μm at a frequency of 1 GHz. Nevertheless, they can prove to be advantageous for the testing of fine cracks.

Fine surface cracks in a metal can be characterized by the way in which they scatter eddy currents induced in it by electromagnetic radiation. Bahr (1981) was able to relate the dimensions of simulated surface cracks, in the form of slots, to the amplitudes and phases of scattered radiation, as would be measured by a conventional microwave receiver located at a fixed distance above the surface. His theory was confirmed experimentally by measurements conducted at a frequency of 98 GHz for slots of different depths and machined by an electrodischarge method.

Studies have been made on the use of a ferromagnetic resonant microwave (FRM) probe (section 7.3.2) for characterizing tight fatigue cracks (Auld, 1981; Auld and Winslow, 1981) having depths of as little as 0.05 mm. Auld and Winslow achieved this high degree of sensitivity by exciting the probe at a resonance frequency of 1 GHz with a $\mathcal{Q}$ factor of 1000.

## 6.7 FIBRE-REINFORCED PLASTICS

Fibre-reinforced plastics consist of strong uniaxial fibres usually a few micrometres in diameter and with high elastic moduli. they are embedded in thin sheets, thickness 0.125–0.25 mm, of an epoxy resin having a low elastic modulus. They are bonded together to provide multilaminates, thickness up to more than 30 mm in a similar manner to plywood; the fibres in adjoining laminates are usually oriented in different directions. Their densities are much lower than those of light alloys but they possess tensile strengths comparable with those of high-quality steel. This makes them very suitable for structural components in aircraft. At present, the most important of these materials are carbon-fibre reinforced plastics (CFRP) and glass-fibre reinforced plastics (GRP), although boron-fibre reinforced plastics have been widely used in the past.

Fibre-reinforced plastics have previously been tested mainly by ultrasonic, radiological or thermal methods. Gros (1996) has observed that the disadvantages of these methods are their high attenuation coefficients in plastic materials. However, there has been much recent interest in the use of eddy currents when the embedded fibres are made from an electrically conducting material such as carbon.

Gros and Lowden (1995) have pointed out some advantages of the eddy current method for testing composite materials: its relative cheapness, its high accuracy and sensitivity, its portability and repeatability, automated operation in real time, its suitability for on-site and in-service inspection, and there is no need for any contact with a sensor.

Here eddy currents are induced in the conducting fibres by the magnetic field originating from the eddy current probe then they pass through the dielectric composite. The magnetic field from the probe has a comparatively low attenuation coefficient; it induces currents to flow through the fibres, giving rise to reverse magnetic fields which have sufficient strength to produce measurable changes in the impedance of the probe.

The fibres in the composite are oriented more or less parallel to one another; in effect, they produce a large number of electrical capacitances in series, capacitances which act as paths for induced eddy currents as well as providing electrical impedances. However, during the manufacture of carbon-reinforced composites, there is usually no need to ensure the fibres remain more than roughly parallel to one another and there are high probabilities of lateral contacts between them, which allows the flow of eddy currents through the fibres (Figure 6.12). This gives rise to anisotropic effects because, for a given input, the current flowing in the effective direction of fibre orientation is higher than in the plane at right angles to it. However, with multilaminates it is usual for the orientations of adjacent sheets to differ by 90°, so as to reduce the overall anisotropy.

Gros (1996) has also observed that certain electrical insulators, such as rubber, can become effectively conducting as a result of the presence of carbon black, thus allowing eddy currents to be used in testing for defects within vehicle tyres.

The value of the standard penetration depth $\delta$ in the composite is given by equation (2.90), which shows how the corresponding frequency $f$ is equal to $1/(\pi\mu\sigma\delta^2)$. Because the material is non-ferromagnetic, the value of magnetic permeability $\mu$ is equal to $4\pi \times 10^{-7}$ H m$^{-1}$, and using the approximation $4\pi^2 \simeq 40$, we have

$$f = 10^7/4\pi^2\sigma\delta^2 \simeq 2.5 \times 10^5/\sigma\delta^2$$

Gros (1996) indicates that, for carbon-fibre epoxy resins, the approximate values of electrical resistivity vary from 50 to 200 μΩ m, from which we get approximate values of electrical conductivity $\sigma$ in the range 5000–20 000 S m$^{-1}$, thus the corresponding approximate frequency range is from $12.5/\delta^2$ to $50/\delta^2$, i.e. 12.5–50 MHz for a standard penetration depth of 1 mm. With the high degree of probe sensitivity now available, especially with the use of shielded coils, one can obtain an effective penetration depth of 6 mm and, with separate transmitting and receiving probes located on opposite sides of the test sample, a thickness of 12 mm can be tested. Thicker samples can be tested using a lower frequency. For example, depending on the value of the electrical conductivity, a frequency of about 6–25 MHz would allow a fourfold increase in thickness. Care must be taken over the choice of frequency to avoid any resonance effects in the sample and to ensure the phase difference is as great as possible between impedance vectors due to changes in lift-off and conductivity.

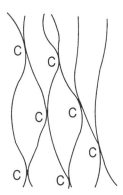

**Figure 6.12** Typical structure of carbon fibres in plastic sheet subjected to eddy current testing: C = the contact point. (Reprinted, by permission, from Lange et al., 1994)

# Neural networks

The use of a high frequency offers two advantages:

- There is a decrease in the capacitative impedance $1/j\omega C$ resulting from the parallel conducting fibres.
- There is an increase in directivity.

Finally, as pointed out by Lange and Mook (1994), great care should be taken to avoid any disturbances caused by variations in probe lift-off during scanning.

## 6.8 NEURAL NETWORKS

A significant recent development in the field of non-destructive testing is the use of the **neural network**, a computing device able to relate the output of a testing instrument to specified properties of a test object. During eddy current scanning of a metal surface, it is possible to measure the depth and, if required, to locate the position of any crack which may be present. Neural networks can be used to process signals from all NDT instruments having numerical outputs.

A detailed discussion of neural networks is not feasible here; the interested reader is referred to Aleksander and Morton (1990). A fairly large number of papers have recently been published on its applications to non-destructive testing, of which those by Kirk and Lewcock (1995) and Charlton (1993) are recommended; Charlton is especially concerned with eddy current testing.

Several types of neural network are available, but apparently most suitable for NDT is the **multilayer perceptron**. The design and operation of neural networks are analogous to those of the brain, whose fundamental unit of information processing is the **neuron**. Neurons are arranged in several layers called **perceptrons**, i.e. input and output layers plus one or more hidden layers; Figure 6.13

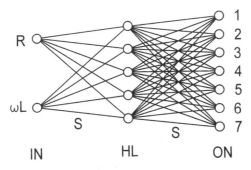

**Figure 6.13** Neural network for eddy current crack detection: HL = hidden layer, IN = input neurons, ON = output neurons, R = input resistance, $\omega L$ = input inductive impedance.

illustrates a network consisting of three layers. Connections between neurons in neighbouring perceptrons are called **synapses**.

The operation of a neural network may be simply explained by considering the surface scanning of a metal object using an eddy current probe to detect cracks and to measure their depths. The inductive and resistive components of impedance, $\omega L$ and $R$ respectively, preferably normalized (equation 4.13a), are taken from the output of the impedance bridge and fed to the appropriate input neurons (IN). The signals then pass through the synapses to the hidden layer (HL). The values of these signals, i.e. the synapse weights, can be adjusted to obtain optimum performance. On reaching the neurons in the hidden layer, the sum of the weighted signals is given a bias in the form of an increase by an amount having a fixed initial value but subject to weighting. The final quantity is then subjected to a change having the form $f(x) = 1/(1 + e^{-x})$, roughly similar in appearance to a magnetic hysteresis curve (Figure 2.13), i.e. near linear for intermediate values and near horizontal for extreme values. This provides the output of a neuron in the hidden layer. The process is repeated to obtain the outputs of the neurons in the next layer, either the next hidden layer or the output layer. The outputs of the neurons in the outer layer are then fed to an appropriate indicator such as a digital display.

To perform properly, the neural network requires suitable values of weighting and bias; it has to be **trained**. To measure the depth of surface cracks in metals, using a surface-scanning eddy current coil, requires the preparation of a suitable calibration block consisting of vertical saw-cuts of different depths in a metal sample having electrical and magnetic properties identical to those of the material under test. For example, if there are six cuts having depths of 0.5–3.0 mm at intervals of 0.5 mm, the neuron outputs from 1 to 6 (Figure 6.13) should indicate crack depths covering this range; output 7 would indicate the absence of any crack.

With the probe in position above a given slit, the weighting is varied during back propagation until the optimum signal amplitude appears at the appropriate output neuron. The process is repeated until each of the calibration slits have been scanned, as well as a 'non-defective' part of the block. Hence, when scanning a metal object made of the same material as the test block, an indication of the crack depth is obtained. The entire procedure can be achieved with the aid of a computer software package specially designed for this purpose; for example, Molyneaux et al. (1995) used a system called Aspirin/MIGRAINES when assessing concrete structures with RADAR using neural networks. To allow for changes in the properties of the coil, normalized values of impedance may be used. On completion of this training, the device is ready for use.

## 6.9 DEFECT IMAGING

### 6.9.1 General considerations

With conventional eddy current testing, defects in a metal are detected and characterized by changes of resistive and inductive components of the impedance of the detecting coil, and the locations of these defects are determined from the position of that coil. Furthermore, visual images of detected defects can be obtained in the form of a display on the screen of a cathode-ray oscilloscope, or perhaps a printout on paper. Formerly it was possible only to acquire an indication of the position and size of a defect, and perhaps some idea of its shape. Following developments in computer technology, one can now obtain displays of detailed variations in geometry of individual defects, using a fairly rapid rate of scan. Thus, the depths of different parts of a crack can be indicated by a printout of the outline of the whole crack in different shades of grey, or perhaps different colours. A brief but highly informative account of this topic has been provided by Guettinger *et al.* (1993).

### 6.9.2 Magneto-optic eddy current imaging (MOI)

When testing over a very large area, a disadvantage of using the conventional procedure is the length of time taken to perform a surface scan using a probe; the probe has to be only a few millimetres in diameter in order to achieve a degree of resolution sufficient for the observation and imaging of very small defects. This would most certainly be the case when testing the skins of 'aging aircraft', rapidly growing numbers of passenger aircraft which have been in service for about 20 years or considerably longer. It is essential these skins are regularly tested for fatigue cracks and corrosion. Another time-consuming factor when using conventional eddy current equipment is the need to strip any paint, which would provide unwanted lift-off; the item has to be repainted after testing.

**Magneto-optic eddy current imaging (MOI)** (Fitzpatrick *et al.* 1993, 1996; Simms, 1993) reduces the testing time and is more sensitive than conventional eddy current techniques; any paint stripping to decrease the amount of lift-off is usually unnecessary. Because the effective probe area is of the order of 6000 mm$^2$, as compared with only a few square millimetres for a conventional defect-scanning probe, very large surface areas can be rapidly examined.

The method makes use of the Faraday effect, in which a beam of polarized light is rotated when it passes through a magnetic field (e.g. Jenkins and White, 1976); the value of the speed of rotation depends on the magnetic field strength and the physical properties

182    *More advanced eddy current testing methods*

of the transparent medium through which the beam travels. In principle the degree of angular rotation of an optical beam, resulting from a magnetic field generated by eddy currents in a metal, can therefore be used to evaluate the amplitudes of the eddy currents. This allows the electrical properties of the metal to be determined along with the location and size of any defects that are present.

With MOI testing (Figure 6.14), a beam of polarized light is directed to a sensor (S). Fitzpatrick *et al.* (1993) used a sensor having a film of bismuth-doped iron garnet, approximate thickness of 3 µm, grown on a substrate of gadolinium gallium garnet, 75 mm in diameter and 0.5 mm thick. According to Simms (1993), this device can provide a rotation of the polarized beam of up to 3000° per millimetre of film thickness, depending on the composition of the film and the crystal structure. The beam is reflected at the surface of a thin copper foil (F) to pass through the sensor for a second time, thereby experiencing further rotation before it is received by an analyser located in a suitable position. This doubles the sensitivity of detection. Eddy currents are induced in the foil (F) by the coil (C), designed so that parallel alternating currents, called **sheet currents**, flow in the area lying directly beneath the sensor. A similar pattern of currents is consequently induced in the test sample.

Any defect at the surface or subsurface would interfere with the pattern of flow of the eddy currents in the test sample, hence in the copper foil, producing a change in magnetic field strength. The change in magnetic field strength creates a change in angle of rotation of the polarized beam, as indicated by the analyser; the value of this change is proportional to changes in magnetic field. The output of the analyser is fed to an imaging device (section 6.9.1).

The frequencies used for the instrument chosen by Fitzpatrick *et al.* were in the range 1.6–104.4 kHz. The sensitivity was considerably

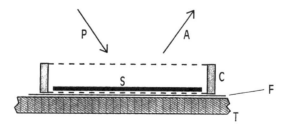

**Figure 6.14** Schematic to show a magneto-optic instrument for eddy current measurements: P = incident polarized light beam, C = coil, S = sensor, F = thin copper foil, T = test sample, A = emergent polarized beam received by the analyser. (Reprinted from Fitzpatrick *et al.*, 1993, by permission of the American Society for Nondestructive Testing)

greater than obtained with more conventional eddy current devices, thus achieving greater penetration and higher degrees of lift-off. Higher lift-off is advantageous in that surface paint thicknesses, perhaps about 0.4 mm, have insignificant effects on image quality.

# CHAPTER 7

# Microwave methods

## 7.1 INTRODUCTION

The electromagnetic microwave method of testing is less commonly used than other electrical NDT methods because of its comparatively limited range of applications and, except in its very basic form, its somewhat greater complexity of instrumentation. However, with the present trend in industry towards the replacement of metals by non-metallic materials and the increase in availability and the corresponding decrease in price of the more sophisticated electronic devices, the use of microwave testing should become more widespread. The microwave method is mainly used for testing dielectric materials in bulk and the surface and subsurface regions of metals. It has the advantage, shared with other electromagnetic techniques, that no contact is necessary between the probe and the object being tested. It is often possible to locate the probe at some distance from the object, which facilitates testing at high temperatures and also the inspection of moving objects.

Due to its comparatively small range of non-destructive testing applications, the extent of the treatment devoted here to microwave testing is fairly limited and fuller details are provided by Dean and Kerridge (1970), Rollwitz (1973), Bahr (1983), Botsco, *et al.* (1986), Botsco and McMaster (1996) and Zoughi (1995). The reader is also referred to standard works (e.g. Adam, 1969; Bailey, 1989) for more general information on microwave technology. A summary of the basic theory of electromagnetic waves in given in section 2.8.

Electromagnetic microwaves are generated at frequencies lying within the band extending roughly from 3 to 300 GHz with corresponding wavelengths *in vacuo* ranging from 100 down to 1 mm, i.e. between, but overlapping with, radio and infrared bands. The microwave band is subdivided into narrower bands, characterized by letters of the alphabet (Table 7.1). Radiation from microwaves can cause a health hazard but the risks are negligible at the low power levels, generally well below 250 mW, sufficient for

## Introduction

**Table 7.1** Principal wavebands within the microwave range

| Band | L | S | X | K | Q | V | W |
|---|---|---|---|---|---|---|---|
| Approximate range (GHz) | 0.4–1.5 | 1.5–5.2 | 5.2–11 | 11–36 | 36–46 | 46–56 | 56–100 |

non-destructive testing purposes. The American National Standards Safety Institute has recommended limits of 10–50 W m$^{-2}$ with a corresponding linear rise in frequency of 0.3–1.5 GHz, and 50 W m$^{-2}$ for frequencies of 1.5–100 GHz (Botsco et al. 1986).

Dielectric materials which are suitable for microwave testing include ceramics, glasses, plastics, fibre-reinforced plastics, concrete, propellant substances, explosives and rubbers. Their mechanical, structural, physical and chemical properties are related to their electrical permittivities and dielectric losses, which can be evaluated by measuring the speeds and attenuations of microwaves passing through them (section 2.8.1). Attenuation can also result from the presence of defects in a material.

Applications of microwave testing include determinations of thickness, degree of porosity, hardness, density, moisture content, anisotropy, chemical composition, degree of aging and presence of defects such as voids, metallic inclusions, large cracks and delaminations. Because of their relatively long range in air at the lower frequencies, microwaves can be used with advantage for performing tests on large installations, including determinations of moisture in coal and grains of cereals moving on conveyor belts, and also detecting buried metal objects. Although microwave methods have proved very effective for locating delaminations in honeycomb structures and fibre composites, they are not generally suitable for detecting thinner cracks for the reason indicated below.

The testing of electrical conductors is restricted to the reflection method for dimension measurements and the application of microwave eddy currents to investigate their surface properties.

Microwave testing can be considered as a possible alternative to ultrasonic testing for electrical non-conducting materials (e.g. Krautkrämer, 1983). Its main advantage is the existence of relatively high transmission coefficients for electromagnetic waves across the boundaries between solid objects and air. This contrasts with the low coefficients for ultrasonic testing, where coupling materials are required. Coupling materials are unnecessary with microwave testing. Furthermore, attenuation coefficients for dielectrics are generally considerably lower with microwaves than with ultrasound. For example, when waves cross an air–solid boundary, the transmission loss for microwaves (section 7.2.2) is usually only a few decibels, but for ultrasound it extends from 30 to 40 dB. At least four boundaries are crossed by waves when passing from the transmitter through the test object to the receiver and the total transmission loss for air-coupled ultrasound is sufficiently high to exceed the dynamic range of most commercially available equipment when conventional ultrasonic probes are used. The low transmission losses of microwaves thus enable the testing of dielectric solids at high temperatures, where ultrasonic coupling fluids become ineffective. Because of the comparatively low attenuation coefficients in air, the microwave transmitters and receivers can be remote from a hot object while suffering only a small extra loss in power. Consider this in relation to the piezoelectric effect, which allows the operation of ultrasonic probes. The piezoelectric effect diminishes with increasing temperature and finally vanishes at the Curie point, which usually lies within the range 100–600 °C, depending on the nature of the transducer used as the probe.

Another advantage of the microwave method over the ultrasonic technique is the considerably greater ease of obtaining large frequency variations at constant amplitude.

The ease of coupling experienced with the microwave testing of non-metals becomes a disadvantage for detecting cracks and other localized defects. In contrast to ultrasonic propagation, the reflection coefficient for electromagnetic radiation at a crack is usually small, partly because of the good coupling with the void inside it and also because its thickness is generally small compared with the wavelength. The microwave method is thus not generally suitable for detecting thin cracks in dielectric materials.

## 7.2 MICROWAVE RADIATION

### 7.2.1 General considerations

Electromagnetic microwaves can be generated by an electronic oscillator and propagated in a coherent manner with plane polarization. Propagation is defined here as taking place in the $z$-direction of a Cartesian system with the electric and magnetic field vectors polarized in the $x-y$ plane and acting in the $x$ and $y$ directions, respectively, i.e. $E_x$ and $H_y$. Either vector can be considered, because they are interdependent, but the use of the electric vector is more convenient since electrical rather than magnetic quantities are usually measured.

The ratio $E/H$ of the electric field vector $E$ to its corresponding magnetic field vector $H$ is known as the intrinsic impedance $\eta$ which, for unattenuated progressive plane waves, is equal to $(\mu/\varepsilon)^{1/2}$, where $\mu$ is the magnetic permeability and $\varepsilon$ is the electrical permittivity. This can easily be shown by differentiating the solutions to the wave equations for $E_x$ and $H_y$, i.e. in the form $\exp j(\omega t - kz)$ (equation 2.72), and substituting them into Maxwell's equations (equations 2.21 and 2.24). It can be seen from the dimensions of $E$ and $H$ that the unit of $\eta$ is the ohm. For free space, $\mu = \mu_0 = 4\pi \times 10^{-7}$ H m$^{-1}$ and $\varepsilon = \varepsilon_0 = (1/36\pi) \times 10^{-9}$ F m$^{-1}$, so $\eta = 376.99\ \Omega$. Values of $\eta$ are very high for metals, where $\varepsilon$ is negligible; this is to be expected because of the low degree of penetration of electric and magnetic fields in these materials. For attenuating media, $\eta$ is complex but the imaginary component is usually small.

Expressions for the speed $c$ and the attenuation coefficient $\alpha$ for dielectric and conducting materials are given in section 2.8 and, for convenience, they are reproduced below.

For dielectrics

$$c^2 = 2/\{\mu\varepsilon'[1 + (1 + \varepsilon''^2/\varepsilon'^2)^{1/2}]\} \tag{7.1a}$$

$$\alpha^2 = \omega^2\mu\varepsilon''^2/2\varepsilon'[1 + (1 + \varepsilon''^2/\varepsilon'^2)^{1/2}] \tag{7.1b}$$

where $\omega$ is the angular frequency and $\varepsilon'$ and $\varepsilon''$ are the real and imaginary components, respectively, of $\varepsilon$. With low attenuation, $\varepsilon''$ tends to zero, and the expression for $c$ reverts to its constant value of $(1/\mu\varepsilon)^{1/2}$ (equation 2.70), otherwise both $c$ and $\alpha$ are functions of frequency.

For conductors

$$c = (2\omega/\mu\sigma)^{1/2} \tag{7.1c}$$

$$\alpha = (\omega\mu\sigma/2)^{1/2} \tag{7.1d}$$

where $\sigma$ is the electrical conductivity.

Attenuation in dielectrics is discussed in section 7.2.5. In electrical conductors, attenuation is caused by the induction of eddy currents which produce magnetic fields in opposition to the applied fields (Chapter 4).

### 7.2.2 Reflection and transmission at a single boundary

#### Normal incidence

Consider plane waves striking a plane boundary between two semi-infinite dielectric media 1 and 2 (Figure 7.1a) at normal incidence and let $E_i$, $E_r$ and $E_t$ be the amplitudes of the electric field at the boundary. We then have

$$r = E_r/E_i \quad \text{and} \quad t = E_t/E_i \tag{7.2}$$

where $r$ is the amplitude reflection coefficient and $t$ the amplitude transmission coefficient, for a single surface. The power (or intensity)

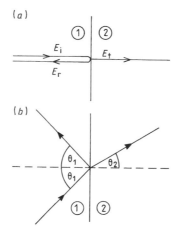

**Figure 7.1** (a) Reflection and transmission of a microwave beam for normal incidence at a plane boundary between two semi-infinite dielectric media, 1 and 2. $E_i$, $E_r$ and $E_t$ indicate electric field amplitudes for incident, reflected and transmitted waves at the boundaries. The reflected beams are superposed on the incident beam but, for clarity, they are shown with a lateral displacement. (b) Reflection and transmission of a microwave beam incident at an oblique angle to a plane boundary between two semi-infinite dielectric media, 1 and 2. The angles of incidence and reflection are $\theta_1$ and the angle of refraction is $\theta_2$.

reflection and transmission coefficients, $R$ and $T$ respectively, are equal to the squares of the corresponding amplitude coefficients, i.e. $R = |r|^2$ and $T = t^2$. The significance of the modulus sign for $r$ is discussed later in this section and in section 7.2.3. If $c_1$ and $c_2$ are respectively the speeds of plane waves in media 1 and 2, we define $n_1$ and $n_2$ as the corresponding refractive indices, defined as

$$n_1 = 1/c_1 \quad \text{and} \quad n_2 = 1/c_2 \tag{7.3}$$

It can be shown (e.g. Born and Wolf, 1970) that

$$r = (n_2 - n_1)/(n_1 + n_2) \tag{7.4a}$$

and

$$t = 2n_1/(n_1 + n_2) \tag{7.4b}$$

Equation (7.4a) shows that when $c_1 > c_2$, $n_2 > n_1$, e.g. for waves in air striking a solid boundary, $r$ takes a positive sign, but with a reversal in direction caused by reflection, $r$ becomes negative, i.e. there is a phase change of 180° of the electric field vector on reflection.

With very few exceptions, e.g. poorly conducting ferrites, the magnetic permeability for a dielectric is virtually the same as for free space, i.e. $\mu_0$. Hence the value of the speed $c$ of electromagnetic waves is inversely proportional to $\varepsilon^{1/2}$ (equation 2.70), where $\varepsilon$ is the electrical permittivity, which is a real quantity for a non-dispersive, hence non-attenuating medium. Tests are usually conducted with the sample in air, where $\varepsilon$ is virtually equal to $\varepsilon_0$, so the ratio of the values of $\varepsilon$ for the media at a boundary is the relative permittivity, or dielectric constant, $\varepsilon_r$. The ratios $c_2/c_1$, hence $n_1/n_2$, are thus equal to $\varepsilon_r^{1/2}$. Equation (7.4a) can therefore be written as

$$r = (\varepsilon_r^{1/2} - 1)/(\varepsilon_r^{1/2} + 1) \qquad (7.5)$$

Most commonly used dielectric materials (Table 2.1) have $\varepsilon_r$ values of between 2 and 5; and from equation (7.5) the corresponding reflection coefficients are 0.17–0.38 (i.e. from $-15$ to $-8$ dB). When one of the media is a metal, the reflection coefficient is very close to unity.

### Oblique incidence

Considering again a plane boundary between two semi-infinite media (Figure 7.1b) but now with incidence at an oblique angle $\theta_1$ with the normal to the boundary, reflection takes place at the same angle with the normal $\theta_1$, but on the other side of it, and the transmitted waves are refracted at an angle $\theta_2$ given by Snell's law, i.e.

$$c_1/\sin\theta_1 = c_2/\sin\theta_2 \quad \text{or} \quad n_1\sin\theta_1 = n_2\sin\theta_2 \qquad (7.6)$$

The values of the transmission and reflection coefficients of these obliquely directed waves depend on the polarization of the electric field vector as well as the angle of incidence and the refractive indices of the media, as expressed by Fresnel's equations (Born and Wolf, 1970):

$$R^{\parallel} = \tan^2(\theta_1 - \theta_2)/\tan^2(\theta_1 + \theta_2) \qquad (7.7a)$$

$$R^{\perp} = \sin^2(\theta_1 - \theta_2)/\sin^2(\theta_1 + \theta_2) \qquad (7.7b)$$

$$T^{\parallel} = \sin 2\theta_1 \sin 2\theta_2/[\sin^2(\theta_1 + \theta_2)\cos^2(\theta_1 - \theta_2)] \qquad (7.7c)$$

$$T^{\perp} = \sin 2\theta_1 \sin 2\theta_2/\sin^2(\theta_1 + \theta_2) \qquad (7.7d)$$

where the superscripts $\parallel$ and $\perp$ indicate that the electric field vectors are respectively parallel and perpendicular to the plane of incidence.

For the special case when the reflected and transmitted beams are perpendicular to one another, i.e. $\theta_1 + \theta_2 = 90°$ so $\tan(\theta_1 + \theta_2) = \infty$, equation (7.7a) shows that $R^{\parallel} = 0$. The angle $\theta_1$ is then equal to $\theta_B$, i.e. the Brewster angle, and equation (7.6) indicates that

$$\tan\theta_B = n_2/n_1 \qquad (7.8)$$

### 7.2.3 Stationary waves

When reflection takes place with normal incidence at a plane boundary between two media (Figure 7.1a), i.e. in the z-direction, interference occurs between the incident and reflected waves, and stationary or standing waves are formed. The pattern of the stationary waves is related to the value of $r$ for a given frequency $f = 2\pi/\omega$. Assuming the attenuation in the medium of incidence is negligible, i.e. the medium is almost lossless, the wave equations may be written as

$$E_1 = E_i \sin(\omega t - kz) \text{ for incident waves} \quad (7.9a)$$

$$E_2 = E_r \sin(\omega t + kz) \text{ for reflected waves} \quad (7.9b)$$

where $k$, the wavenumber, is equal to $2\pi/\lambda$. There is no advantage here in expressing the phase angles in the exponential form. The resultant electric field strength $E$ is given by

$$E = E_1 + E_2 = E_i \sin(\omega t - kz) + rE_i \sin(\omega t + kz) \quad (7.10)$$

where $r$ is the amplitude reflection coefficient and is given by $r = E_r/E_i$. The value of $E$ depends on whether $r$ is positive or negative, according to the relative values of $n_1$ and $n_2$ (equation 7.4a). If medium 2 is a metal, $r$ is virtually equal to unity, i.e. $E_1 = E_2$.

**When $n_2$ is greater than $n_1$**, e.g. when medium 1 is air and medium 2 is a solid or liquid, $r$ is positive and thus equal to $+|r|$, so equation (7.10) becomes

$$E = 2|r|E_i \cos kz \sin \omega t + (1 - |r|)E_i \sin(\omega t - kz) \quad (7.11a)$$

**When $n_2$ is less than $n_1$**, $r = -|r|$ and

$$E = 2|r|E_i \sin kz \cos \omega t + (1 - |r|)E_i \sin(\omega t - kz) \quad (7.11b)$$

In the first term on the right-hand side of each of these equations the amplitude varies sinusoidally with $z$, so the maximum and minimum amplitudes are each located in fixed positions at half-wavelength intervals with a distance of separation between a maximum and its neighbouring minimum equal to one quarter-wavelength; hence the term 'stationary waves'. Figure 7.2 shows the variations of amplitude with $z$ for values of $r$ having magnitudes equal to 0.2, 0.6 and 1, respectively. The ratio of maximum amplitude $E_{max}$ to minimum amplitude $E_{min}$ is defined as the stationary (or standing) wave ratio SWR, expressed as

$$\text{SWR} = (1 + |r|)/(1 - |r|) \quad (7.12)$$

When $|r|$ is equal to unity, e.g. when medium 2 is a metal, $E_{max}$ is equal to $2E_i$ and $E_{min}$ to zero; then the positions of the maxima and minima are called nodes and antinodes, respectively. When $n_2$ is greater than $n_1$, the maximum amplitude occurs at the boundary (where $z = 0$); when $n_2$ is less than $n_1$, there is a minimum amplitude at the boundary.

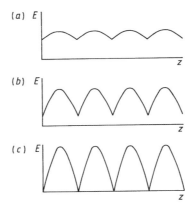

**Figure 7.2** Variation of electric field amplitude $E$ with distance $z$ for stationary waves arising from reflection at normal incidence at a plane boundary between two semi-infinite media for reflection coefficients (a) $r = 0.2$, (b) $r = 0.6$ and (c) $r = 1$. The curves are predicted using equations (7.11) and all are drawn on the same scale. The origin of each graph does not necessarily indicate the position of the boundary.

Consider the formation of stationary waves as a result of internal multiple reflections at the opposite surfaces (i.e. where $z = 0$ and $z = d$, respectively) of a parallel-sided dielectric object of refractive index $n_2$ and located between two semi-infinite dielectric media having refractive indices $n_1$ and $n_3$, respectively (Figure 7.3). Half-wavelength resonance occurs when $d$ is equal to an integral number $n$ of half-wavelengths, i.e. $n\lambda/2$, when either $n_1 < n_2 > n_3$ or $n_1 > n_2 < n_3$. In the first instance, the amplitudes of the electric fields at both boundaries of the object are maximum, and in the second instance they are minimum. Quarter-wavelength resonance takes place when $d$ is equal to $(2n - 1)\lambda/4$ and either $n_1 < n_2 < n_3$ or $n_1 > n_2 > n_3$, thus giving a maximum amplitude at one boundary and a minimum at the other.

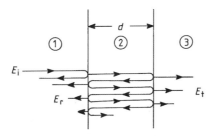

**Figure 7.3** Reflection and transmission of a microwave beam for normal incidence at a parallel-sided dielectric slab (2), thickness $d$, between two semi-infinite dielectric media (1, 3). Read the remarks in the caption to Figure 7.1a.

192                    *Microwave methods*

The treatment can be extended to three dimensions for its application to the closed cavity resonator in the form of a rectangular metal box containing air or, where appropriate, a vacuum. If the dimensions of the box are $X \times Y \times Z$ in the $x$, $y$ and $z$ directions respectively, and a wave is propagated in a given direction, the characteristics of the wave (i.e. $E$ or $H$) can be resolved in the $x$, $y$ and $z$ directions and the conditions for resonance are

$$[(n_x/X)^2 + (n_y/Y)^2 + (n_z/Z)^2]^{1/2} = 2/\lambda \qquad (7.13a)$$

with the resonance frequencies $f$ given by

$$f = (c/2)[(n_x/X)^2 + (n_y/Y)^2 + (n_z/Z)^2]^{1/2} \qquad (7.13b)$$

where $n_x$, $n_y$ and $n_z$ are integers, corresponding to the $x$, $y$ and $z$ directions respectively. These resonance frequencies are sometimes called eigentones or eigenfrequencies.

For an open cavity resonator, the open end lies in a plane normal to the direction $z$ of wave motion and half-wavelength resonance occurs in the $x$–$y$ planes. Quarter-wavelength resonance takes place along the $z$-direction. This device becomes effectively a closed cavity resonator if the open end is covered with any material being tested.

The resonance frequency of a cylindrical cavity resonator having its axis in the $z$-direction depends only on two dimensions, i.e. $Z$ and the radius $R$, because of axial symmetry.

### 7.2.4 Reflection and transmission with two parallel boundaries

Consider again normal incidence at a lossless parallel-sided dielectric medium of thickness $d$ and a refractive index $n_2$ (Figure 7.3) separating two dielectric media having refractive indices $n_1$ and $n_3$. Multiple reflections take place between the two boundaries and stationary waves build up (section 7.2.3). The resultant power reflection coefficient $R'$ at the first boundary (Born and Wolf, 1970) is given by

$$R' = (r_{12}^2 + r_{23}^2 + 2r_{12}r_{23} \cos 2kd)/(1 + r_{12}^2 r_{23}^2 + 2r_{12}r_{23} \cos 2kd)$$
(7.14a)

where $k = 2\pi/\lambda$ and $\lambda$ is the wavelength in medium 2, and $r_{12}$ and $r_{23}$ are the amplitude reflection coefficients for the boundaries between the media 1 and 2 and 2 and 3, respectively, when considered in isolation. Taking the special case of microwaves incident normally at a parallel-sided dielectric slab of thickness $d$ located in air, $r_{12}$ and $r_{23}$ have equal magnitudes but opposite signs (equation 7.4a), i.e. $r_{12} = r$ and $r_{23} = -r$. Making the substitution $\cos 2(kd) = 1 - 2\sin^2(kd)$, we have

$$R' = 4r^2 \sin^2(kd)/[(1 - r^2)^2 + 4r^2 \sin^2(kd)] \qquad (7.14b)$$

## Microwave radiation

The corresponding power transmission coefficient $T'$ for the slab is clearly equal to $1 - R'$ and the amplitude coefficients $t'$ and $r'$ are equal to the square roots of the respective power coefficients.

For a given frequency, $R'$ and $r'$ vary with the thickness $d$ of medium 3 and reduce to zero when $\sin kd = 0$, i.e. when $d$ is equal to an integral number of half-wavelengths. The reflection coefficients reach their maxima when $\sin kd = 1$, i.e. when $d$ is equal to an odd number of quarter-wavelengths. $R'$ and $r'$ tend to zero when the ratio $\lambda/d$ becomes very small. If the reflection and transmission coefficients are expressed in decibels, there is no need to differentiate between the power and amplitude coefficients (section 2.5.6).

In the derivation of equation (7.14a) it is assumed the attenuation in medium 2 is zero and the value of the electric field $E$ varies as $\sin(\omega t - kx)$ for progressive waves. When the attenuation is finite, $E$ then varies as $\exp(-\alpha z)\sin(\omega t - kz)$, where $\alpha$ is the attenuation coefficient (section 2.8). Thus it is necessary to multiply the expressions in equation (7.14a) containing $\cos 2kd$ by $\exp(-2\alpha d)$. Equation (7.14b) then becomes

$$R' = 2r^2\{1 + \exp(-2\alpha d)[2\sin^2(kd) - 1]\}/\{1 + r^4 + 2r^2\exp(-2\alpha d) \\ \times [2\sin^2(kd) - 1]\} \tag{7.14c}$$

To facilitate normalization, this equation can be expressed as

$$R' = 2r^2\{1 + \exp(-2d\delta/\lambda)[2\sin^2(2\pi d/\lambda) - 1]\}/\{1 + r^4 \\ + 2r^2\exp(-2d\delta/\lambda)[2\sin^2(2\pi d/\lambda) - 1]\} \tag{7.14d}$$

where $\delta = \alpha\lambda$ is the logarithmic decrement (section 2.5.2), i.e. the attenuation per cycle or wavelength. Figure 7.4 shows how the reflection coefficient varies with the ratio $d/\lambda$, i.e. thickness/wavelength, for values up to unity, in accordance with equation (7.14d). Here we take a typical value of $r = 0.2$; this corresponds to a boundary between a dielectric and air, in which the respective speeds $c$ of the microwaves are equal to $2 \times 10^8$ m s$^{-1}$ and $3 \times 10^8$ m s$^{-1}$ (equations 7.3 and 7.4a). The reflection coefficients are expressed in decibels (dB), for reasons given earlier. The effects of increasing attenuation increase the minimum values of the wave amplitude (and intensity) and introduce asymmetry in the standing-wave curves in the dielectric.

Equation (7.14b) is also applicable to a parallel-sided air gap in a dielectric material. This is of special interest because, when its thickness $d$ is small, it resembles a planar crack. If $d \ll \lambda$, $\sin kd$ approximates to $kd$ in equation (7.14b) and if

$$(kd)^2 \ll (1 - r^2)^2/4r^2$$

$$R' \approx 4r^2(kd)^2/(1 - r^2)^2 \tag{7.15}$$

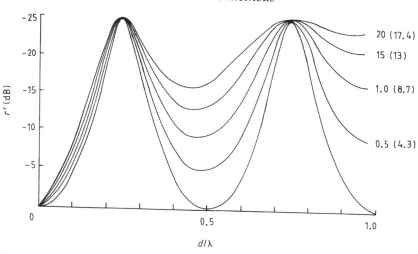

**Figure 7.4** Variation of reflection coefficient r' (dB) with the ratio $d/\lambda$ for normal incidence at a parallel-sided slab of dielectric material in air when $r = 0.2$ (equation 7.14d) for different attenuations per wavelength given in terms of the log. dec. $\delta$ and the corresponding decibel value (in parentheses). The slab thickness is $d$ and $\lambda$ is the wavelength within the slab.

From equation (7.5) we see that

$$R' \approx \pi^2(\varepsilon_r^{1/2} - 1/\varepsilon_r^{1/2})^2(d/\lambda)^2 \tag{7.16a}$$

i.e.

$$r' \approx \pi(\varepsilon_r^{1/2} - 1/\varepsilon_r^{1/2})(d/\lambda) \tag{7.16b}$$

where $\varepsilon_r$ is the relative permittivity of the solid dielectric. Equation (7.16b) shows that when $d \ll \lambda$ the amplitude of the reflected waves is proportional to the thickness of the gap.

When the gap takes the form of a fine crack, typically 0.01 mm thick, in a Perspex (Lucite or Plexiglas) sample the application of a high frequency is needed to produce a suitable value of $d/\lambda$. For example, at a frequency of 200 GHz when $\varepsilon_r = 2.6$ and $\lambda = 1$ mm, $r'$ becomes equal to 0.009, which corresponds to a reflection loss of 40 dB, and a weak signal is received. The reflection coefficient increases with the width of the gap and any consequent resonance can be minimized by increasing the frequency bandwidth so as to reduce the $\mathcal{Q}$ factor (equation 2.47) to a suitably low level.

### 7.2.5 Attenuation of microwaves

It is shown in section 2.8.1 how the attenuation coefficient for a dielectric material is related to the electrical permittivity $\varepsilon$ when expressed as a complex quantity, i.e. $\varepsilon' - j\varepsilon''$. The value of $\varepsilon''$ depends on what is called the dielectric heating of the material caused by the alternating electric field giving rise to changes in the polarization of the molecules, hence producing displacement currents (section 2.4.2). The attenuation coefficient relating to this phenomenon, i.e. the absorption coefficient, depends on the frequency and also the temperature of the given material. This is a relaxation effect which is characterized by the absorption coefficient increasing with frequency and attaining a maximum value at what is called the relaxation frequency $f_r$ then decreasing with further increase in frequency. When $f \ll f_r$ the absorption coefficient increases as the square of the frequency. The phenomenon is accompanied by a change in wave speed, i.e. velocity dispersion. Relaxation absorption is the principal cause of attenuation in water; this phenomenon enables the assessment of moisture content in dielectric materials (Botsco and McMaster, 1996).

Another cause of attenuation in dielectric materials is Rayleigh scattering, which may arise from distributed inclusions such as pores and metal particles. Provided the wavelength $\lambda$ is at least one order of magnitude greater than the average linear dimension $D$ of the scattering object (i.e. $D > 10\lambda$), the amount of energy scattered is proportional to $D^3/\lambda^4$ (e.g. Jenkins and White, 1976). The attenuation coefficient $\alpha$ is then given by

$$\alpha = KD^3 f^4 \qquad (7.17)$$

where $f$ is the frequency and $K$ a constant that depends on the properties of the dielectric material, the scattering objects and their mean sizes along with their distribution density. Scattering takes place uniformly in all directions.

## 7.3  MICROWAVE INSTRUMENTATION

### 7.3.1  General considerations

In its basic form the microwave apparatus required for non-destructive testing consists of a source connected through a waveguide to a transmitting horn, which propagates electromagnetic radiation into the object under test. A separate receiving horn, linked by a waveguide to a detector connected to an output device, picks up the radiation after it has been either transmitted, reflected or scattered, depending on the location of the horn. The output device indicates the required characteristics of the electric field, e.g. amplitude and phase.

Figure 7.5a shows an arrangement of the basic equipment comprising two horns, one transmitting and the other receiving, positioned for through-transmission measurements, i.e. aligned axially to one

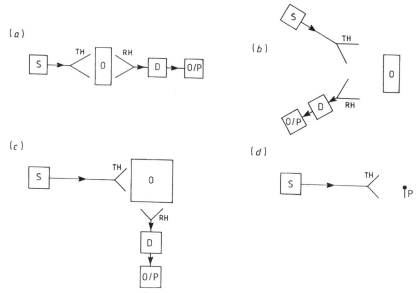

**Figure 7.5** Basic arrangements for microwave testing: (a) dielectric transmission, (b) reflection at oblique incidence, (c) orthogonal scattering, (d) in a stationary wave (SW) field with insertion of a probe (P). S = source, TH = transmitting horn, RH = receiving horn, O = test object, D = detector, O/P = output device. Straight lines represent waveguides.

another on opposite sides of the object. The two-horn method can also be used for oblique incidence and, for this purpose, the axes of the horns are coplanar and directed at equal angles on either side of the normal at the surface (Figure 7.5b) to enable the receiving horn to pick up the reflected radiation. Figure 7.5c indicates the positions of the horns for scattering measurements, where the portion of the beam scattered at right angles to the direction of incidence is accepted by the receiving horn, i.e. orthogonal scattering is detected. Figure 7.5d illustrates a simple but somewhat crude method of using a detector in the form of a probe mounted on a short post, made from a dielectric material, to bring it to the level of the axis of the transmitting horn. The probe scans the stationary waves between the transmitting horn and the reflecting surface of the object being tested. The standing wave ratio and the reflection coefficient (equation 7.12) are obtained by measuring the output of the probe at the positions of maximum and minimum amplitudes. The precision of this method depends on the relative dimensions of the probe (about 2 or 3 mm diameter) and the wavelength. An error of less than 5% is obtained at the upper practical frequency limit of about 1 GHz, for which the wavelength in air is 30 mm.

This basic two-horn system is inexpensive, simple to assemble and simple to operate, but it has limited precision. The sensitivity of the method can be considerably improved by providing a reference signal direct from the source to the detector, allowing comparisons for both amplitude and phase (see later). When a single horn is used to receive reflections resulting from normal incidence, a reference signal is essential (section 7.3.5).

### 7.3.2 Microwave sources and detectors

The main requirement for a microwave source suitable for non-destructive testing is a low energy output consistent with safety and with the linearity and stability of the radiation characteristics. This requirement can be satisfied without difficulty by using one of the conventional sources that are readily available and described in standard texts (e.g. Bailey, 1989; Adam, 1969).

Until recently, klystron tubes were the main sources of low-amplitude microwaves. A klystron can exist in one of several forms and is a vacuum tube in which electron beams are velocity modulated to excite a cavity into resonance. The resonance frequency can be varied by changing the volume of the cavity with a tuning screw. Klystrons suffer the disadvantages common to all thermionic vacuum tubes but they are cheap, easily obtainable and reliable, and by suitable choice it is possible to obtain a klystron which can operate at low power and at any frequency within the microwave spectrum.

The reflex klystron (Figure 7.6) is perhaps the most suitable of the klystrons for microwave testing. It generates a beam of electrons which is accelerated through holes (H) in the metal wall of a tuned parallel-sided and electrically earthed closed cavity (C). The beam emerges through similar holes in the opposite wall and is then retarded and subsequently reversed in direction by a negatively charged electrode acting as a repeller (R). This produces a cyclical variation of electron velocity, whose frequency depends on the value of the electrical potential difference between the cavity walls and the repeller; the potential difference can be adjusted to allow the cavity to resonate. The natural frequency of the cavity can be varied within a limited range by using the tuning screw (T).

Other solid-state devices, e.g. the Gunn diode, are simpler and have greater elegance, compactness and versatility. They have largely superseded klystrons as microwave sources for NDT purposes. Although they are limited in power for many microwave uses, their outputs are perfectly adequate for NDT applications. The upper frequency limit attainable with these devices is currently about 100 GHz, but improvements are very likely to increase this figure considerably in the near future.

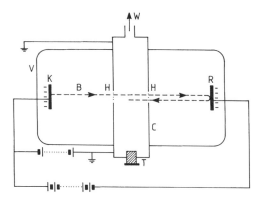

**Figure 7.6** Schematic of the reflex klystron: B = electron beam, C = tuned cavity, H = holes, K = cathode, R = repeller, T = tuning screw, V = vacuum tube, W = waveguide coupler.

The Gunn diode operates on the principle of the transferred electron effect and consists of a thin layer of gallium arsenide (GaAs), across the opposite faces of which is applied a direct electric field. With increase in field strength from zero, the carrier drift velocity increases linearly to a maximum for a field of about 320 V mm$^{-1}$, after which there is a steady velocity decrease with increasing field strength. Conditions become unstable at this optimum field, where the conduction electrons are rapidly redistributed to minimize the total energy; consequently, regions of high and low electric fields are formed, bounded by layers of charge. These layers are mobile and oscillate at microwave frequencies which can be controlled by electrical or mechanical tuning. Although mechanical tuning can be achieved by changing the volume of a resonant cavity using a rotating screw, the Gunn diode can be tuned electronically over a very wide range of frequencies in a variable manner, thus making it highly suitable as a frequency-modulated source.

The use of microwave detectors presents few difficulties. The simplest form is a silicon diode connected to the receiving horn by a waveguide; alternatively the detector can be located at any required point in the field (Figure 7.5d). Silicon diode detectors are highly sensitive, and for the low field levels used in NDT, the output current is proportional to the input voltage, hence the magnitude $E$ of the electric field. Thermistors also act as sensitive detectors, but being thermally activated, they measure power rather than voltage. Compared with diode detectors, they also respond less quickly to radiation. Mention should be made here of the ferromagnetic resonance probe (FMR) is a resonant device which can be located close to the object being tested, allowing its state of resonance to be affected by the properties of that object (Figure 7.7a). An FMR probe suitable for

## Microwave instrumentation

microwave operation consists of a small sphere a fraction of a millimetre in diameter and made from a ferromagnetic crystal such as yttrium iron garnet (YIG). The sphere is encircled by a wire loop and lies in a direct magnetic field provided by a direct current passing through a coil or by a permanent magnet. On applying the EMF to the loop, the magnetic moment induced in the sphere precesses about an axis parallel to the field. The frequency of the precession depends only on the strength of this field which, by suitable adjustment, can produce resonance. If the microwave field strength varies as a result of changes in the properties of the object being tested, there is a departure from resonance, hence a change in the impedance of the wire loop.

Auld and Winslow (1981) located an FMR probe at a point only just above a metal surface. The probe was made to resonate at a frequency of 1 GHz with a $Q$ factor of 1000. The magnetic field was supplied by a suitably located samarium–cobalt permanent magnet several millimetres long and located very close to the probe (Figure 7.7b), thus inducing eddy currents in the metal (section 6.4).

### 7.3.3 Waveguides

Waveguides provide a highly effective means of conveying microwaves from the source to the horn and from the horn to the receiver;

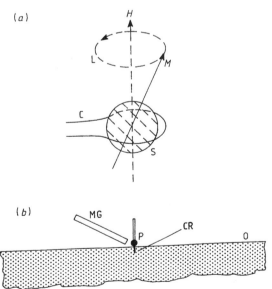

**Figure 7.7** Ferromagnetic resonance (FMR) probe: (a) schematic, (b) positions of probe and magnet with respect to a cracked metal object. S = YIG sphere, C = coil, H = applied direct magnetic field, L = locus of precession for magnetization $M$ of the sphere, MG = magnet, P = probe and holder, O = metal test object, CR = crack.

they are essential for frequencies above 1 GHz. Waveguides are ducts having either circular or rectangular cross-sections, usually rectangular, and are made from rigid and accurately dimensioned conducting material. Sections of waveguides can be bolted together at end-flanges to form a circuit and junctions can be made by suitably designed T-pieces. Various accessories (section 7.3.5) can be inserted, as required.

The waves progress through a waveguide in a zigzag manner, with lateral reflections at opposite surfaces, and they are characterized by an impedance $Z_{TE} = E/H$; the subscript TE indicates the electric field rather than the magnetic field is used here to characterize the radiation. When the cross-section is unbounded or when its dimensions are very large compared with the wavelength, $Z_{TE}$ is equal to the intrinsic impedance $\eta$ (section 7.2.1). However, with propagation in waveguides, the value of $Z_{TE}$ is governed by both lateral dimensions and wavelength. When there is an abrupt change in cross-section of the waveguide, perhaps arising from an open end in contact with free space, there is a correspondingly abrupt change in impedance. Partial reflection therefore occurs in the axial direction. This effect, however, is minimized by using a horn, which allows a gradual change in impedance along its length, enabling good impedance matching between the waveguide and the propagation medium. The result is that any reflection at the mouth of the horn is negligible.

### 7.3.4 Horns

The radiation initially emerges from the horn as a parallel beam, called the near-field or Fresnel zone. Divergence then takes place in the far-field or Fraunhöfer zone with the wave intensity decreasing in accordance with the inverse square law, i.e. the amplitude decreases in inverse proportion to distance from the mouth of the horn. The length $l$ of the near field for a circular aperture is approximately given by

$$l = D^2/4\lambda \tag{7.18a}$$

where $D$ is the diameter and $\lambda$ the wavelength. For an aperture with a rectangular cross-section, Botsco et al. (1986) quoted the length of the near field as

$$l = A^2/2\lambda \tag{7.18b}$$

where $A$ is the largest dimension of the rectangle.

In common with ultrasonic testing, it is advisable to ensure the microwave path length lies entirely in either the near field or the far field of the transmitting horn when measurements of amplitude are made. Near-field measurements do not require corrections for beam spreading but complications arise with the formation of stationary waves, except in materials where the attenuation is sufficiently high for the total path length to be less than the length of the near field. The decrease of amplitude in inverse proportion to the distance from the transmitting horn produces an attenuation of 6 dB per doubling of path length, and appropriate corrections should be made to the measured amplitudes and powers.

At frequencies below 1 GHz the length of the near field is usually small enough for the radiation to be considered entirely within the far field, for practical purposes at least. Considering a rectangular horn with the larger cross-sectional dimension equal to 100 mm and propagating into air ($c = 3 \times 10^8$ m s$^{-1}$) at a frequency of 1 GHz, equation (7.18b) indicates a length of the near field equal to only 16 mm. On the other hand, at a frequency of 100 GHz, it is equal to 1.6 m.

### 7.3.5 Microwave circuits

For the reasons given in section 7.4.1, microwave measurements for non-destructive testing are usually confined to continuous waves. Figure 7.8 shows the typical arrangements of the components for microwave measurements at a single frequency, showing both single- and two-horn arrangements. The guided waves pass from the source (S) to the T-junction (J), between which there may be an amplitude modulator (M). The advantage of amplitude modulation, usually at a

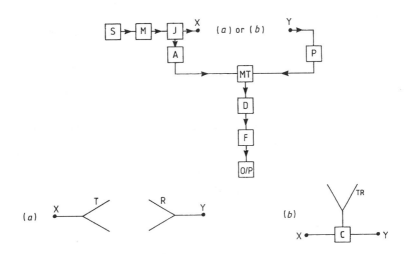

**Figure 7.8** Simplified block diagram for a microwave NDT circuit showing alternative arrangements for (a) transmission and (b) reflection testing. S = source, M = amplitude modulator, J = T-junction, A = attenuator, MT = magic tee, P = phase shifter, D = detector, F = filter, O/P = output device, T = transmitting horn, R = receiving horn, TR = transmitting/receiving horn, C = direction coupler.

kilohertz frequency, is that when the signals arrive at the output stage of the circuit, the microwave frequency can be filtered out to leave a noise-free low-frequency signal, thereby achieving greater ease and accuracy of detection.

The T-junction (J) acts as a beam splitter and thus provides a way to compare the phase and amplitude of the received signal with those of a reference signal supplied directly from the source. The main part of the beam continues to the point X, and for two-horn operation it passes through X to the transmitting horn (T); the receiving horn (R) is connected to Y. For single-horn operation, a direction coupler (C) lies between X and Y, designed to prevent any radiation bypassing the horn. The received radiation is channelled from Y to a phase shifter (P) which can operate either mechanically or electronically, e.g. using a silicon PIN diode. A typical example of a mechanical phase shifter is a rotating vane made from a lossless dielectric material and crossed by the radiation. The path length through the vane, related to the change in phase, depends on its angle of rotation. The phase of the received signal is varied until it coincides with the phase of the reference signal; the amount of change can be read from a suitably placed circular scale.

The reference signal from the T-junction (J) then passes through an attenuator (A) which provides a continuous adjustment of the wave amplitude, again using either a PIN diode or a mechanical device. An example of the latter is a suitably mounted rotating thin film metal vane in which the attenuation coefficient is very high compared with that for dielectric materials. A change in path length through the vane would produce a corresponding change in attenuation; the angle of rotation depends on the change of attenuation, similar to the rotating-vane phase shifter.

The reference and received signals are then combined by a device known as a magic tee (MT) which equalizes the powers of two signals and produces a 180° phase difference between them (Chatterjee, 1988) to provide a signal of zero amplitude. This is a null method in which only variations from the reference signal are measured, thus enhancing the sensitivity. The output from the magic tee then passes to the detector (D) and finally, for modulated waves, through a filter (F) to a suitable output indicator (O/P) such as a meter, cathode-ray oscilloscope or a computer-aided device.

## 7.4 MICROWAVE MEASUREMENTS

### 7.4.1 General considerations

The quantities usually measured with microwave testing are amplitude, frequency and time delay (or phase difference); the measurements can be performed with high degrees of precision using modern equipment. This has been made possible by rapid developments made in other aspects of microwave technology, e.g. communications, in which there is widespread interest.

Microwave methods of non-destructive testing are almost invariably conducted with continuous waves, instead of pulses, although pulse methods are used for long-range applications such as radar. However, the comparatively long pulses which are easily generated are unsuitable for locating discontinuities such as boundaries and defects in small dielectric objects. Short pulse lengths of the order of 1 mm, as used with ultrasonic testing, are necessary to obtain a suitable degree of resolution, but 1 mm microwave pulses are only of the order of 5 ps in duration and are therefore difficult or impossible to process with today's commercial equipment. Highly sophisticated devices able to reach these standards may well be available in specialized research laboratories but the expense of obtaining, maintaining and operating them is usually prohibitive if they are required only for routine non-destructive testing.

Fortunately, however, the swept-frequency continuous wave method, which provides output signals in the shape of half-waves having very short durations, is proving a useful substitute for the conventional pulse method when detecting discontinuities in dielectric objects (section 7.4.4). At present, relatively simple fixed-frequency continuous wave methods (sections 7.4.2 and 7.4.3) are commonly used for measuring velocities and thicknesses. Continuous standing wave and cavity resonance methods are often highly suitable for attenuation measurements.

When making microwave measurements of amplitude and attenuation, the importance of allowing for beam spreading in the far field of the transmitting horn (section 7.3.4) should be remembered. Where possible, any correction can be avoided by the choice of a frequency compatible with a long near field.

### 7.4.2 Fixed-frequency continuous progressive wave methods

The simplest methods for microwave testing make use of continuous waves at a fixed frequency, but they are effective only for the following:

- Normal incidence at an object having either two plane parallel sides or a single plane side, provided there is sufficient attenuation to prevent the formation of detectable standing waves.
- Oblique incidence at a plane surface of an object.
- Amplitude measurements of radiation scattered at an angle to the direction of propagation.

### Testing highly attenuating materials

When testing parallel-sided dielectric objects, it is generally important to ensure that radiation reflected from the far surface should fall below the noise level of the measuring instrument. This may entail reducing the instrument's sensitivity, perhaps to an unacceptably low level. With a single transmitting/receiving horn (Figure 7.8b) the wave velocity, the refractive index or the electrical permittivity can be determined from the reflection coefficient $r$ (equation 7.4a) in one of two ways. One is to compare the amplitude of the reflected beam with the amplitude of a reference signal supplied by the source. The other is to compare the amplitude of the beams reflected from the surface of the test object and a metal sheet placed in the same position after the test object has been removed. The value of $r$ for the test object is readily determined because $r$ is virtually equal to unity for the metal sheet.

The reflection method can also be applied to measuring the thickness of a parallel-sided slab or sheet made from metal, or other very highly attenuating material, with the two transmitter/receiver horns located on opposite sides (Figure 7.8a) and excited in phase with one another. In the absence of the sample, waves from each transducer travel in opposite directions from one another, each beam having the same path length $s$ between the horns. On interposing the slab or sheet of thickness $d$ with one of the surfaces at a distance $z$ from one of the horns, the respective path lengths become $2z$ and $2(s - z - d)$, where $s$ is the separation of the horns; the path difference is thus equal to $2(s - 2z - d)$ with the corresponding phase difference $\delta\phi$ given by

$$\delta\phi = 2k(s - 2z - d) \qquad (7.19a)$$

where $k = \omega/c_0$, $\omega$ is the angular frequency and $c_0$ the speed of electromagnetic waves in air. If the phase of one of the reflected beams is then changed by 180°, $z$ is eliminated and

$$\delta\phi = 2k(d - s) \qquad (7.19b)$$

A knowledge of the exact location of the test object between the horns is thus irrelevant, provided the opposite surfaces are perpendicular to the beams.

The use of two horns, one transmitting and the other receiving, located on opposite sides of a parallel-sided block of a highly attenuating dielectric material (Figure 7.8a) enables the measurement of velocity, attenuation coefficient and thickness, provided the level of the signal is sufficiently high to allow penetration of the incident waves but low enough to prevent the detection of reflected waves. If the velocity and attenuation coefficient are to be determined, the thickness $d$ of the sample should be known. On the other hand, if it is necessary to evaluate $d$, a knowledge of the velocity $c$ in the object is required. The phase change $\delta\phi$ brought about by the introduction of the test object is related to the time delay $\delta t$ as follows:

$$\delta\phi = \omega \delta t = \omega d(1/c_0 - 1/c) \tag{7.20}$$

The attenuation coefficient $\alpha$ is evaluated simply by comparing the amplitude $A$ of the received signal when the sample is present with the amplitude $A_0$ when it is absent

$$A = A_0 \exp(-\alpha d) \tag{7.21}$$

where $\alpha$ is expressed in neper per metre. Alternatively $\alpha$ can be expressed as $20 \log_{10}(A/A_0)$ dB m$^{-1}$, where 1 neper is equal to $20 \log_{10} e$ dB, i.e. approximately 8.7 dB. Care should be taken to ensure the object and the receiving horn lie entirely within the near field or the far field of the transmitting horn (section 7.3.4). A correction is required when object and horn lie entirely within the far field.

### Continuous wave testing at oblique incidence

When continuous microwaves are incident obliquely at a plane boundary between two dielectric media (Figure 7.1b), the angles of reflection and incidence are equal to one another, and the relationship between the angles of incidence and reflection depends on the microwave speeds in the media (equation 7.6). The reflection and transmission coefficients depend on these angles and the relationships for polarizations of the electric field vectors parallel and perpendicular to the plane of incidence are given by equations (7.7). Provided the thickness of the material relative to the width of the beam is sufficiently large and provided the attenuation is high enough to prevent interference from reflections at the lateral boundaries, the formation of unwanted stationary waves is avoided.

Of particular interest is the special case when the incident waves are polarized parallel to the plane of incidence at the Brewster angle $\theta_B$ (equation 7.8), when the reflection coefficient is equal to zero. The transmitting and receiving horns are mounted on a goniometer, a device used for X-ray spectroscopy (McMaster, 1963f) and ultrasonic reflection measurements (Blitz, 1971), which allows their axes to rotate so they are inclined equally to the normal for all angles of incidence. The horns are rotated, each with increasing angle of incidence $\theta_1$, until the received signal disappears, i.e. when $\theta_1 = \theta_B$, and the ratio of the refractive indices of the two media is equal to $\tan \theta_B$. With air as the medium of incidence, the speed of the waves and, hence, the electrical permittivity of the material under test are readily determined. The degree of accuracy depends principally on the quality of the goniometer.

### Amplitudes of scattered waves

The degree of scattering caused by distributed defects in a dielectric sample has successfully been measured using the technique in Figure 7.5c, illustrated in its basic form, where the axis of the receiving horn is perpendicular to the axis of the transmitter. In practice, a calibration is made with samples in which the sizes of the scattering sources are known along with their distribution. The choice of frequencies is optimized from equation (7.17), which shows how the amplitude of the scattered radiation is proportional to the cube of the mean linear dimension of the scattering objects.

### 7.4.3 Fixed-frequency continuous stationary wave method

The fixed-frequency stationary wave method can be used for dielectric materials to measure wave velocities, thicknesses, degrees of anisotropy, positions of discontinuities (such as abrupt structural changes and delaminations) and attenuation, which can be related to the chemical, structural and physical properties of the material concerned.

When continuous waves at a fixed frequency are incident normally at the surface of a parallel-sided dielectric sample, reflections occur at both the opposite surfaces and stationary waves are formed between these surfaces. The characteristics of these stationary waves are related to the reflection coefficient, hence they are related to the transmission coefficient (equations 7.14 and Figure 7.4). Although both reflection and transmission techniques (Figure 7.8) can be used for testing dielectric materials, reflection techniques are often preferred because access to only one surface is necessary. An increase in sensitivity of reflection measurements can be achieved by placing a plant metal reflector in close contact with the rear surface to preclude through-transmission.

### Thickness measurements

Thickness measurements on dielectric materials having uniform composition and structure (e.g. constant electrical permittivity) can be achieved by comparing either the amplitude or phase of the received reflected or transmitted waves with a reference datum supplied directly from the source (Figure 7.8). The choice of frequency depends on the thickness to be measured, and for amplitude comparisons, especially when determining variations of thickness from those of a standard sample, it is important to avoid the regions of maxima and minima on the curve of the amplitude reflection coefficient (Figure 7.4). The maximum change in thickness which can be determined without any ambiguity, when measuring changes in amplitude, is equal to one quarter-wavelength, but when observing changes in phase it increases to one half-wavelength. The method can be used for absolute measurements of thickness if this quantity is approximately known.

When the thickness of a parallel-sided object is small compared with the wavelength, equations (7.16) show how the amplitude of the reflected or transmitted waves is directly proportional to the thickness, provided the attenuation coefficient is sufficiently small to allow through-penetration. For a sample having a dielectric constant of 2.5 and a thickness of 10 mm, an error of about 3% is incurred for a frequency of 1 GHz if this linearity is assumed.

When a parallel-sided slab of dielectric, thickness $d$, is placed in close contact with a metal surface, total reflection takes place at the metal surface and a mirror image is formed, effectively increasing the thickness to $2d$, which can be evaluated from the reflection coefficient. The phenomenon has been applied to measuring thicknesses of 25–50 µm in polythene films deposited on aluminium foil with reflection measurements at a frequency of 9.4 GHz (Bahr, 1983). Since there is no through-penetration of electromagnetic waves in the metal foil at this frequency, the measurements are not affected by variations in the thickness of the foil, as might be found when using eddy current methods.

## *Measurements of refractive index (or electrical permittivity)*

The technique described above for measuring the thicknesses of parallel-sided dielectric slabs, where the magnetic permeability is equal to $\mu_0$ can be used for determining values of wave velocity $c$, hence refractive index $n$ and electrical permittivity $\varepsilon$, provided the thickness is known, since $c = 1/n = 1/(\mu_0 \varepsilon)^{1/2}$. The conditions for the choice of frequency are the same as for thickness determinations. Measurements made in different specified directions can be used to determine degrees of anisotropy in a dielectric material but the method of polarimetry (section 7.4.5) may be more convenient for this purpose.

## *Measurements of attenuation*

Attenuation coefficients can be determined by measuring reflection coefficients and using either equation (7.14c) or (7.14d). Figure 7.4 shows the effects of attenuation on the relationship between the reflection coefficient, as measured in decibels, and the ratio of thickness to wavelength. The optimum frequency for measurement is clearly the frequency where the thickness of the test object is an integral number of half-wavelengths. This value of thickness corresponds to a minimum value of reflection coefficient at which changes in the attenuation coefficient can be determined with the highest degree of sensitivity.

Attenuation measurements are widely carried out to determine moisture content and a comprehensive discussion of this topic is provided by Botsco *et al.* (1986) who list 120 materials and products which have been successfully gauged for moisture using the microwave technique. Another interesting application is the monitoring of cure in resins (Bridge, 1987).

### 7.4.4 Variable-frequency methods

Continuous frequency variations with microwave measurements can be carried out most effectively by modulating the frequency of the source. The Gunn diode is highly suitable for this purpose because it can be tuned electronically over a wide frequency range, often equal to a complete microwave band (Table 7.1). The modulation is usually performed so the frequency varies linearly with time. Two variable-frequency applications are discussed here, i.e. the resonant cavity and the swept-frequency echo (or time-of-flight) methods.

#### *The resonant cavity method*

The resonant cavity microwave technique is normally used for measurement with dielectric materials, although it can be applied to inspect metal surfaces. It uses a conducting cavity (section 7.2.3) having an open end in contact with the material being tested. The cavity is energized by a swept-frequency source, and amplitude peaks can be observed after suitable signal processing, from which the values of the resonance frequency and $\mathcal{Q}$ factor (sections 2.5.2 and 2.8.1) can be determined. When the open end is placed in contact with the test object, these values change by amounts which depend on the properties of both the resonator and the dielectric material under test, i.e. wave velocity and attenuation, hence the real and imaginary components of electrical permittivity, which contribute to the value of the reflection coefficient. The degree of attenuation of the resonating cavity is affected by the properties of the material being tested, including moisture content, the presence of scattering objects, e.g. distributed porosity, and defects such as cracks.

In its simplest form, the device used for this method is a frequency-modulated Gunn diode located inside the cavity. The variation of the impedance of the diode with frequency indicates the values of the resonant frequency and $\mathcal{Q}$ factor, from which the values of the components $\varepsilon'$ and $\varepsilon''$ of $\varepsilon$ can be obtained. King and Latorre (Botsco *et al.* 1986) used a more sophisticated method with a coaxial cavity resonator (Figure 7.9); with the aid of a metal flange at the open end of the resonator, they obtained a spreading of the beam in the surface region of the dielectric material being tested, i.e. the fringing field. The use of this device thus helps to avoid the appearance of standing waves. After amplification and modulation over a frequency range of 30 kHz, the reflected signal was combined with a reference signal fed from the source and suitably adjusted for phase and amplitude. A trace indicating the variation of amplitude with frequency was displayed finally on the screen of a cathode-ray oscilloscope, thus indicating the values of the resonant frequency and $\mathcal{Q}$ factor.

Most applications use open rectangular and cylindrical cavities, which allow deeper penetration; King and Latorre have employed these cavities to detect cracks and delaminations in dielectrics. The presence of these defects brings about changes in the frequency and $\mathcal{Q}$ of resonance. Although this technique makes it difficult to locate defects with any degree of precision, it is possible to monitor the growth of surface-breaking cracks after an initial calibration with artificial cracks consisting of cuts of known depth.

The resonant cavity method can also detect anisotropy in materials, an example is the determination of the direction of cellulose fibre in paper. Tiuri and Liimatainen (Botsco *et al.* 1986) used an open cavity which could resonate at the same frequency for two electric fields polarized at right angles to one another. With the paper placed in position, two different resonant frequencies are detectable with the degree of separation increasing with the anisotropy. In the absence of anisotropy, resonance occurs at only a single frequency.

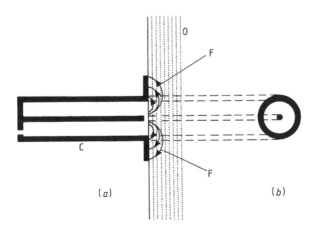

**Figure 7.9** Cavity resonator used by King and Latorre: (a) side elevation, (b) cross-section. C = coaxial resonator, F = fringing field, O = object under test. (Reprinted from Botsco *et al.*, 1986, by permission of the American Society for Nondestructive Testing)

### The swept-frequency continuous wave echo (or time-of-flight) method

The swept-frequency continuous wave echo varies the frequency with time in a linear manner, between the limits $f_a$ and $f_b$, at regular intervals, e.g. several hundred times per second. For a typical application, a beam of microwaves having a frequency $f_0$ at a time $t_0$ is generated from the swept-frequency source (SFS) (Figure 7.10) and passes through a waveguide to a dual-direction coupler (DC) where part of the radiation, acting as a reference signal, is diverted to a mixer (M). The main part of the beam continues to the transmitting/receiving horn (TR) where it is radiated at normal incidence to a parallel-sided dielectric object having a thickness $d$. The beam is partially reflected at both the front and rear surfaces and returns to the horn with a differential time delay $t$ equal to $2d/c$, where $c$ is the speed of the microwaves through the object.

On arrival back at the coupler (DC), the first reflected beam is diverted to the mixer where combination takes place with a reference beam transmitted at a later time $t_1$ when the frequency has increased to $f_1$. The frequency difference $f_1 - f_0$ is proportional to the time delay $t_1 - t_0$. Similarly, a frequency difference $f_2 - f_0$, proportional to a time delay $t_2 - t_0$, is obtained when combination takes place between the reference signal and the second reflected beam. In each case the combination produces sum and difference frequencies; the output is filtered to leave only the difference frequency, which is usually within the mid or upper kilohertz range. Thus for a sweep time of 100 μs, corresponding to a change in frequency from 75 to 125 GHz and a time of say 1 ns for the return trip between the coupler and the nearside reflecting surface of the test object, $f_1 - f_0$ is equal to only 500 kHz, which is easily handled by conventional electronic detection equipment. If the object has a thickness $d$ of 20 mm and its material has a wave velocity of $2 \times 10^8$ m s$^{-1}$, there is a time difference of

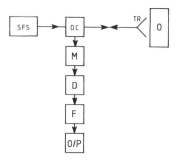

**Figure 7.10** Simplified block diagram for the swept-frequency echo method. SFS = swept-frequency source, DC = dual-direction coupler, M = fixer, F = filter, D = detector, TR = transmitting/receiving horn, O = test object, O/P = output device.

0.2 ns between the two reflections. Thus, $t_2 - t_0 = 1.2$ ns and $f_2 - f_0 = 600$ kHz.

By performing a Fourier transformation, single sine wave (half-cycle) peaks corresponding to each of these echoes appear on the screen of a cathode-ray oscilloscope. The timebase of the oscilloscope is connected across the X-plates and calibrated against distance to indicate the peak separation, hence the thickness of the test object (Figure 7.11). If the sample contains a defect such as large crack, delamination or disbond, oriented roughly at right angles to the direction of the beam, the location of the defect is indicated by another peak lying between those corresponding to the two surface echoes. It should be noted how the position of a reflection is indicated by the apex of the peak, not by the leading edge, as experienced with the ultrasonic pulse technique (Krautkrämer, 1983).

Since stationary waves are not formed in the object being tested, the method is highly suitable for thickness gauging and also for defect detection, although defect detection is subject to the restrictions discussed in section 7.1 and illustrated by equation (7.16). Any limitations of this technique when applied to dielectric materials should be compared with the greater limitations imposed by ultrasound. As with ultrasonic pulse–echo testing, imaging is possible with B and C scanning (Krautkrämer, 1983). The possibility of imaging by microwave tomography is discussed by Voskreseneskii and Voronin (1987).

The method is also suitable for attenuation measurements, which can be made by comparing the heights of the front and back echoes, after allowing for reflection losses and any far-field beam spread. It is important here to select the range of the frequency sweep and the rate of change of frequency with time in a manner which allows compatibility with the rate of variation of attenuation coefficient with frequency.

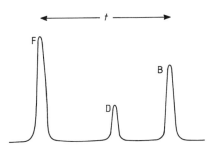

**Figure 7.11** Cathode-ray oscilloscope display of echoes from front (F) and back (B) surfaces of a test object using the swept-frequency echo method. D is an echo corresponding to a defect and $t$ is the time delay between the surface echoes.

### 7.4.5 Polarimetry measurements

In section 7.4.4 it was shown how the resonant cavity method can be applied to investigate anisotropy in dielectrics. However, it is more usual to use the polarimetry technique, whereby the microwave detector is rotated in the plane of its cross-section, and to note any change in the amplitude of the electric field, which would indicate directions of polarization caused by anisotropy. Consider wave propagation in the $z$-direction with the electric vector $E$ in the $x$-direction. The principal axes in the medium are oriented along the directions $x \sin \theta$ and $x \cos \theta$, where the wave speeds are $c_1$ and $c_2$, respectively. Propagation in the medium then takes place at these two different speeds with corresponding amplitudes $E_x \sin \theta$ and $E_x \cos \theta$, and the resultant polarization is elliptical.

A brief discussion of how this method is used to determine anisotropy in polymers and fibre-reinforced plastics has been given by Rollwitz (1973), who also describes a means of characterizing directions of surface defects in metals from the polarizations of back-scattered reflections. More recent work on the subject is described by Konev *et al.* (1989), who use the term 'microwave ellipsometry' as an alternative title.

### 7.4.6 Testing with RADAR

The methods of microwave testing described in this chapter have proved highly suitable for comparatively small-scale work. But when examining bigger objects, the use of RADAR has been advantageous, especially for slabs and structures of reinforced concrete; this is because concrete has a low attenuation coefficient at lower gigahertz frequencies and very high reflection coefficients at the surfaces of reinforcing steel bars. Furthermore, RADAR equipment is comparatively easy to manipulate. With small-scale microwave equipment, the main topic of this chapter, horns are normally used as probes, but on the exceptional occasions when care has to be taken to minimize changes in the field pattern, detectors may be employed in the form of small spheres having a diameter of a millimetre or so (section 7.3.2).

In RADAR terminology, **antenna** is used to describe both a source and a detector. The choice of antenna depends on the application. Besides those mentioned above, there are many different types:

- A slit oriented laterally in the direction of either the magnetic or electric vector.
- A small aperture to provide a point source located at the focal point of a hyperbolic reflector and directed towards it, so that a parallel beam is emitted in a forward direction.

Because of the greater distances generally involved with RADAR testing, it becomes feasible to make use of pulsed waves. Chen *et al.* (1994) report the use of RADAR to test concrete blocks, both with and without steel reinforcement, having dimensions 610 mm × 610 mm × 150 mm and containing a variety of simulated defects characteristic of the concrete used in constructing the decks of bridges. The operating frequency was about 1 GHz with a pulse repetition frequency of 5 MHz. A single wide-band antenna was used, acting as both transmitter and receiver. The RADAR signals reflected back from the test samples were converted to audio signals by a sampling process.

Identifying the signals received from defects, either simulated or real can be very difficult. Chen *et al.* (1994) used a display of colour profiles for this purpose. However, Molyneaux *et al.* (1995) at Liverpool University made use of a neural network (section 6.8) in a preliminary investigation testing concrete for voids and steel reinforcements. Artificial voids and steel bars of different sizes and depths were put into a tank containing a specially developed oil–water emulsion having electromagnetic properties similar to concrete. The network was trained by repeatedly scanning each of the 153 artificial defects.

The reader who is interested in the principles, techniques and applications of RADAR is recommended to consult either Edge (1993) for a longer account, or Scanlon (1987) for a shorter one. Those who are particularly concerned with concrete testing by RADAR may find it useful to consult two consecutive articles in *Nondestructive Testing and Evaluation*: Padaratz and Forde (1995) cover theoretical aspects and Bungey and Millard (1995) describe experimental studies.

CHAPTER 8

# Miscellaneous methods

## 8.1 GENERAL CONSIDERATIONS

This final chapter covers the remaining significant electrical methods of testing. They are mainly based on simple principles, straightforward in their applications and reliable in their operation. Some of the methods, such as the use of strain gauges, are widely used but receive relatively little attention in the various NDT journals probably because they are so effective there is not much demand for improvement. An important exception is the AC potential drop method, which has undergone rapid development in recent years in view of its ability to provide accurate assessments of surface cracks in metals.

## 8.2 POTENTIAL DROP METHODS

### 8.2.1 General considerations

The basic potential drop method of testing consists of passing a steady current $I$ through a conducting test sample and measuring the resultant potential difference $V$ between two points $C$ and $D$ separated by a distance $s$ in the direction of $I$ and at which a pair of conducting prods is in contact with the surface (Figure 8.1a). The value of $V$ is a function of $I$, $s$, the electrical conductivity (or resistivity), the shape, the size and the physical state of the sample. The method can thus be used to identify properties of the sample such as its nature, thickness, structure and temperature. Because the distance $s$ need be only a few millimetres, the technique can be used to determine spatial variations of electrical conductivity, perhaps caused by a lack of homogeneity or the presence of localized defects such as cracks. Although this method has been used for many years to measure electrical resistivities of semiconductors, only recently has it been applied successfully to the non-destructive testing of metals. This is because the very low values of resistances of metals bring about very small potential differences between the points $C$ and $D$ and also because difficulties arise in establishing good electrical contacts

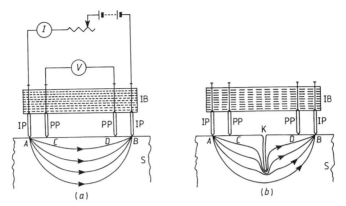

**Figure 8.1** Four-prod probe for resistivity measurements: lines of current flow (a) for a defect-free homogeneous sample and (b) for a sample with a surface crack. IB = insulating probe body, PP = potential prod, IP = current prod, S = sample, K = crack.

between the metal prods and the surfaces. These obstacles have been largely overcome with improved instrumentation, and the method is now becoming more widely used for materials testing and evaluation.

### 8.2.2 The DC potential drop method

An early application of the DC potential drop (DCPD) method was made by Valdes (1954) using semiconductor samples for the assessment of their purity, which varies with electrical conductivity. Valdes used a probe (Figure 8.1a) that consisted of four metal prods, their tips in intimate contact with the material under test. The prods were spring-loaded to enable the exertion of sufficient pressure to maintain good contact. A steady direct current entering the material at $A$ and leaving it at $B$, followed the paths shown in the figure, and the potential difference between $C$ and $D$ was measured. The distance $AB$ should be short enough, e.g. less than 10 mm, to prevent any detectable current from penetrating the lower surface of the test object. This ensures the pattern of the current flow lines is characteristic of the properties and structure of the conductor, not its geometry.

Valdes related the electrical conductivity $\sigma$ to the current $I$, potential difference $V$ and the distances of separations $s_1 = AC$, $s_2 = CD$ and $s_3 = DB$ of the prods, assumed to be in contact with the surface of a semi-infinite medium, i.e.

$$\sigma = (I/2\pi V)[1/s_1 + 1/s_2 - 1/(s_1 + s_2) - 1/(s_2 + s_3)] \qquad (8.1a)$$

If $s_1 = s_2 = s_3 = s$ then

$$\sigma = I/2\pi Vs \qquad (8.1b)$$

Alternatively, the probe is calibrated by placing it on the surface of a standard sample having a known value of conductivity $\sigma'$ and carrying a current $I$ so that a potential difference $V'$ appears between the points $C$ and $D$. The probe is then placed on the surface of the sample to be measured at the same current $I$; a voltage $V$ between $C$ and $D$ is observed. The conductivity $\sigma$ of the material under test is thus equal to $\sigma'V'/V$. To achieve reproducible results, uncertainties arising from intermittent contact resistance must be eliminated, firstly by ensuring the surface is clean then by adjusting the spring-loadings of the prods until $V$ attains a constant value.

The Valdes method produces accurate results for semiconducting materials which have sufficiently low electrical conductivities for any residual contact resistances between the prods and the surface to be negligible compared with the resistance of the material between $C$ and $D$.

This technique has also been applied to steel samples for measuring their thickness and the depth of surface cracks. Inaccuracies may also be introduced by the appearance of thermal EMFs at the contacts of the potential difference prods with the metal surface but they can be eliminated by using prods made from the same material as the test object or by simply repeating the potential difference measurement with the current in the reverse direction then taking the mean of the two values.

The DC potential drop (DCPD) method may be used to measure depths of surface-breaking cracks after they have been detected by another technique, e.g. magnetic particle inspection; it may also be used to size cracks prepared for modelling purposes. It is most effective when the probe is oriented so that the crack comes as close as possible to bisecting the line CD (Figure 8.1b), which indicates the diversion of the current flow lines. Provided that no detectable current reaches the lower surface of the test object, the depth of the crack for a given prod spacing can be related to the ratio $V/V_0$, where $V$ and $V_0$ are the potential differences between $C$ and $D$, firstly over the crack then over a defect-free part of the conductor; the current $I$ is maintained constant throughout. The separations of $A$, $B$, $C$ and $D$ depend on the depth of the crack. Gu and Yu (1990) have shown, that for depths of about 7 mm or less in 30CrMnSi low-alloy steel, suitable separations are $AB = 10$ mm and $CD = 5$ mm, with $AC = 2.5$ mm. They used a numerical modelling technique for calibration and predicted values of $V/V_0$ for crack depths of up to 10 mm with differences from measured values varying by amounts of up to 7.4%. It was concluded that ultrasonic methods are preferable for measuring cracks deeper than 7 mm.

In practice, a crack depth can be measured with the aid of a calibration block made from the same material as the sample being tested and containing fine cuts simulating cracks of known depth. The unknown depth $d$ of the crack in the test sample can then be determined from the relationship between the ratio $V/V_0$ and $d$ given by the calibration curve.

Some degree of success has been achieved with the testing of steel components and structures such as pressure vessels and railway tracks (McMaster, 1963a). The degree of sensitivity increases with the value of the current and, with sufficient care, the method has proven highly reliable. Equation (8.1b) assumes the prods are equally spaced and the current does not penetrate to the lower surface; it shows that to obtain a potential difference output of $5\,\mu V$ for a mild steel defect-free sample ($\sigma \approx 6\,MS\,m^{-1}$) with $s = 5\,mm$, the current should have a value in the region of $1\,A$, assuming the absence of parasitic contact phenomena.

The method is valuable for observing the growth of cracks in metal test pieces under varying loads, as a means of verifying fracture mechanics problems. Because the test rig is semi-permanent, difficulties arising from uncertainties in the prod contacts can be eliminated by soldering them to the surface of the sample, thus making it possible to measure values of potential differences in nanovolts with a high degree of reliability. However, as shown in section 8.2.3, considerable improvements have come with the development of the AC potential drop and field measurements.

When the DCPD method is used to measure the thickness of a metal object, the current flow has to penetrate the lower surface of the object because the thickness is determined by the area through which penetration occurs. Here a greater prod separation is clearly required, typically $12\,mm$ with a corresponding distance between $A$ and $B$ of $36\,mm$. The DCPD method has been applied to measuring corrosion thinning in ship hulls made of $12.5\,mm$ thick mild steel; the reported accuracy was 3% (McMaster, 1963a). More recent work on corrosion detection is reported by Kaup and Santosa (1995).

### 8.2.3 AC potential drop and field methods

The use of alternating currents with the four-prod method for evaluating surface cracks has the advantage of concentrating most of the current in the region just below the surface, i.e. corresponding to the penetration or skin depth $\delta$, as defined in equation (2.90). Dover and Collins (1980) quoted a value of $\delta = 0.13\,mm$ for a mild steel sample having an electrical conductivity of $5.8\,MS\,m^{-1}$ and a relative magnetic permeability of 500 for a frequency of $5\,kHz$. This provides a considerably higher potential difference (e.g. by a factor of several tens) for a given value of current than would DC measurements having a current penetration of several millimetres for the prod separations discussed in section 8.2.2. Further advantages are (a) the thin layer of current follows the profile of a crack, so its depth can be calculated in a simple manner (see below) and (b) contact potentials are eliminated because AC generation produces periodic reversals of current.

The skin effect has an important consequence: there is no need to locate the current prods close to the potential prods, and Figure 8.2 shows the effect of having current contacts remote from the region of interest. It illustrates the paths of current flow and the orthogonal equipotential lines. This technique, aptly named AC field measurement (ACFM), has been described in some detail by Collins et al. (1985). The current contacts are suitably located in fixed positions, about 100 mm on either side of the crack, and within the area of inspection the lines of current flow are parallel to one another and perpendicular to the line of the crack; hence only the potential difference prods need to be moved. As an alternative, when contact between the current leads and the metal sample is not feasible, the current can be induced in the sample by locating several parallel straight wires, equally spaced by an amount sufficiently small to ensure a uniform current density and connected to one another in parallel.

Consider the simple case of a straight vertical crack of uniform depth and extending the whole width of the metal sample. Let $s$ be the separation of the voltage prods, $d$ the depth of the crack and $V_1$ and $V_2$ the respective potential differences with and without the crack.

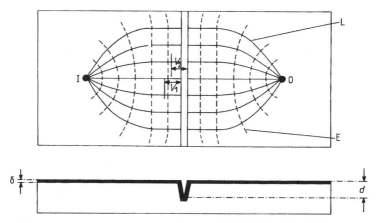

**Figure 8.2** Field distribution for an alternating current in a flat plate conductor containing a simulated surface crack of constant depth $d$. L = lines of current flow, E = equipotentials. The shaded area indicates the penetration zone of depth $\delta$. (Reprinted from Charlesworth and Dover, 1982, by permission of EMAS Ltd)

Because $V_1$ and $V_2$ are each proportional to the current path lengths, respectively $s$ and $s + 2d$, it follows that

$$d = (s/2)(V_2/V_1 - 1) \tag{8.2}$$

Collins *et al.* have also obtained expressions for potential distributions on metal surfaces for simulated cracks in the shape of circular arcs and semi-ellipses. Together with real cracks of different depths, the simulated cracks can be scanned with the potential difference prods straddling the crack and the line joining the prods remaining perpendicular to the crack. Alternatively, an equally spaced array of probes can be located in a stationary position across the crack and multiplexed to provide a high-speed scan without the need to monitor any mechanical motion.

Since the skin effect concentrates the current in a zone within a fraction of a millimetre of the surface, an alternating current of only 1 or 2 A and fed by widely spaced contacts is sufficient to provide a current density over a large area. This allows measurement of potential differences to a high degree of sensitivity, which is advantageous for large structures and components. If the DC method were to be used in this way, a current in the region of several hundred amperes might be required to achieve a degree of voltage sensitivity comparable with the AC method, taking into account the greater accuracies generally obtained with AC instrumentation. Currents of this magnitude are generally unstable and they also produce unwanted heating.

Collins *et al.* (1985) describe an instrument called the ACFM crack microgauge, specifically designed for ACFM measurements, with a current output of 2 A and operating at frequencies in the range 600 Hz to 6 kHz. With its very low noise level, the voltmeter has a resolution of several nanovolts and is carefully screened from the current input stage of the circuit to prevent unwanted signals being picked up. During operation the current and voltage leads should be physically isolated, where possible, to avoid any coupling between them; any parallel leads should be twisted together.

It is important to avoid any induction by the applied current of magnetic fields in the loop formed by potential probes and the surface of the metal; this can largely be achieved in the manner of Figure 8.3. Induction by the probe/metal loop can contribute substantially to the potential difference when testing non-ferromagnetic metals; it increases linearly with frequency, whereas the increase in potential difference that results directly from the current flowing through a resistance is proportional only to the square root of the frequency. The net effect is to set a practical upper limit of about 6 kHz on the operating frequency. However, these two voltages are out of phase with one another by an angle of 45° and the unwanted component can

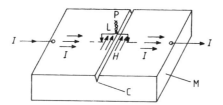

**Figure 8.3** Location of potential difference probe (P) astride a crack (C) in a metal conductor (M), showing induction of alternating magnetic field $H$ by the current. Twisted leads (L) are also illustrated. (Reprinted from Charlesworth and Dover, 1982, by permission of EMAS Ltd)

be reduced, at frequencies below 6 kHz, by using a phase-sensitive detector. Figure 8.4a illustrates the design of a potential difference probe and indicates how the effects of induced voltages are minimized by twisting the leads and by screening.

The ACFM technique has proved highly effective in measuring crack sizes and monitoring the rates of crack growth in a metal sample under stress, and Dover and his colleagues have shown how to acquire much valuable information on the properties of cracks in welds, including welds at pipeline junctions, and in screw threads, aircraft wheels, etc. Stored in a computer, this information can be retrieved when a test is made to characterize the size and growth rate of a newly discovered crack.

The method is highly suitable for underwater measurements and is now increasingly used for everyday non-destructive testing as well as for non-destructive evaluation. It has an advantage over the eddy current method in that it can be used to measure cracks in steels having depths greater than 5 mm and with higher accuracy. Figure 8.4b illustrates a commercial version of the ACFM crack microgauge manufactured by Technical Software Consultants; it can be used for examining welded metal structures. A probe induces a uniform alternating current of 1 A in the test sample and measures the associated electromagnetic fields close to the surface. Excitation can take place at frequencies of 0.2, 1 and 5 kHz. The instrument can test any conducting material. At a frequency of 5 kHz the current remains constant to within 0.5%. One of the portable versions is designed for underwater inspection. All versions can be used with dedicated computer software to eliminate operator error during calibration. The ACFM crack microgauge can also be used for ACPD testing.

Although it overcomes the need for electrical contact, the ACFM technique usually still requires material contact between the probe and the object surface; this is to provide a standard measure of lift-off. The probe arrangement (Figure 8.5) consists of a magnetic yoke excited with a coil carrying an alternating current (e.g. 6 kHz) to provide a uniform field over the region of interest. It also has two

## Potential drop methods

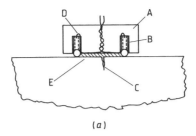

(a)

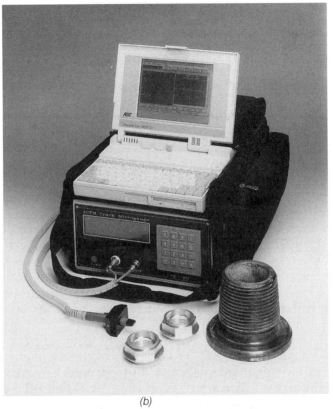

(b)

**Figure 8.4** (a) ACFM probe: A = dielectric probe body, B = ball point fitted into split sleeve, C = fatigue crack, D = connecting wire soldered to the tube and passed down the split, E = minimized pick-up loop area. (b) ACFM crack microgauge. (Part (a) reprinted from Charlesworth and Dover, 1982, by permission of EMAS Ltd; part (b) reproduced by permission of Technical Software Consultants Ltd)

sensors in the form of small coils, each 1 mm in diameter; the coils receive $B_x$ and $B_z$, which respectively represent the components of induced flux density tangential and perpendicular to the surface. The device can be used to measure cracks in screw threads and in T butt welds.

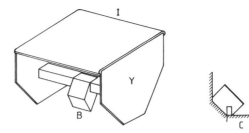

**Figure 8.5** Probe used for ACFM without electrical contact and the manner of its location at the root of a T butt weld: I = inducing coil, Y = yoke, B = $B_x$ and $B_y$ sensors, C = crack. (Reprinted from Collins *et al.*, 1990, by permission of Graham and Trotman and the Society for Underwater Technology)

The components of **B** are required because there are no contact prods to provide a reference scale for length. However, perturbations in the flux density variations during the scan are caused by the contributions from both ends of the crack, and the length scale is given by the relative positions of the maxima and minima of $B_x$.

By measuring the electrical resistivities in concrete, the ACDP method can be used to monitor corrosion in reinforcement bars (Millard *et al.*, 1989). The prods were hollow, spring-loaded and filled with a conductive gel of graphite in ammonia to ensure good electrical contact. They were equally spaced and 50 mm apart, thus forming a Wenner array. The alternating current method was employed mainly to avoid electrolysis at the contacts. Repeatability of measurement with an error of less than 1% was achieved with relatively dry concrete. Optimum conditions were obtained for a current supply of 250 μA at a frequency of 300 Hz and resistivities of about 1.2 kΩ m ($\sigma = 0.83$ mS m$^{-1}$) were measured.

CNS Electronics has produced a resistivity meter using a Wenner four-probe array (Figure 8.6). Produced by a battery, the current is interrupted at regular intervals and thus provides alternating rectangular waves at a frequency of approximately 13 Hz and a constant peak-to-peak voltage of 25 V. The device can measure resistivity values up to a maximum of 2 MΩ cm (i.e. 200 kΩ mm).

In reinforced concrete the measured resistivities are affected by the appearance of double-layer capacitances at the bar–concrete interfaces; this must be taken into account. The capacitances produce resistive components that decrease with increasing frequency (equations 8.7). Adding these components to the true resistance, independent of frequency, gives a total resistance that decreases with increasing frequency. Gowers and Millard (1991) showed how the capacitive effect was negligible at frequencies of below approximately 10 Hz, so any measurements within this range would provide 'true' values of resistivity.

# Resistance strain gauges

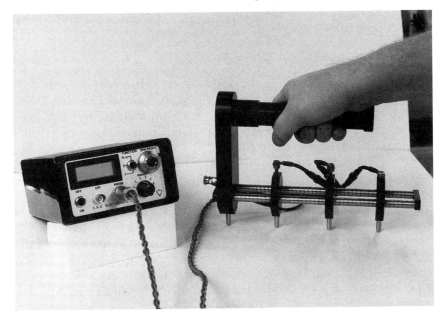

**Figure 8.6** CNS resistivity meter. (Reprinted by permission of CNS Electronics Ltd)

Further work to develop the potential drop and field measurements has been reported by Frise and Sahney (1996), Gowers and Millard (1991), Lugg and Raine (1995) and Topp (1994).

## 8.3 RESISTANCE STRAIN GAUGES

An important non-destructive testing device is the resistance strain gauge (e.g. Perry and Lissner, 1962), which operates on the principle that a conductor's resistance increases with its length (equation 2.1). This device is commonly used to monitor strains in components and structures under stress and can provide an effective early warning against failure. It can also be used for thickness measurements. Strain gauges are employed for objects on the test bench and in service, especially under the critical conditions encountered in the aerospace, petroleum and nuclear industries.

The strain gauge element is a length of electrically conducting wire placed in adhesive contact with the test object so as to indicate either dimensional changes or motion. An increase $\delta l$ in length $l$ is accompanied by a corresponding decrease $\delta A$ in the cross-sectional $A$ in accordance with the value of the Poisson ratio $v$ (e.g. Cottrell, 1964) and it can easily be seen that

$$\delta A/A = (1 - 2v)(\delta l/l) \tag{8.3}$$

and because $v$ cannot exceed 0.5 – in practice it rarely exceeds 0.25 for most conductors – there is a resultant increase in the magnitude of the ratio $l/A$ for a positive strain, hence an increase in resistance.

It is important that the temperature coefficient of resistance of the material of the strain-gauge element is low and that thermoelectric effects are minimized at connections. Elements made from alloys containing nickel, e.g. copper–nickel, nickel–chrome and nickel–iron, usually meet these conditions; the electrical resistivities of these alloys are of the order of $0.01\,\Omega\,\text{m}$ (conductivity $= 100\,\text{S}\,\text{m}^{-1}$).

The configuration of the wires acting as strain-gauge elements can take several forms (McMaster, 1963b), depending on the shape of the surface to which it is fixed, and the adhesive, e.g. cellulose acetate, epoxy resins, ceramics, needs to have properties that are not adversely affected by the stresses producing the strains. Increased sensitivity can be obtained in one of the following ways:

- Arrange the elements parallel to one another and connect them in series to provide a grid containing a long length of wire but covering a small area.
- Wind them round a small sheet of insulating material acting as a card-type of bobbin (Figure 8.7).

The resistance is usually measured by means of a DC Wheatstone bridge (section 2.6.2), allowing the resistances of two strain gauges to be compared, one in contact with the test object and the other acting as an unstrained standard. The strain gauges are in adjacent arms of the

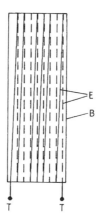

**Figure 8.7** Winding of strain-gauge element (E) around a flat cardboard bobbin (B) to leave terminals (T).

bridge, so the readings of a galvanometer in the detecting arm act as an indication of strain, after suitable calibration.

Because the materials used for strain-gauge elements have relatively high resistivity and exist as thin wires, the strain can be evaluated with a high degree of sensitivity. This can be appreciated from equation (2.1). With an element having a resistance of the order of 1 k$\Omega$, a strain of $10^{-6}$ produces a resistance change of 1 m$\Omega$, which can be measured without difficulty. Although semiconductor strain gauges made, for example, from silicon are considerably more sensitive than metal ones perhaps by a factor of 60, they are highly susceptible to temperature fluctuations, so they are more suitable for use in temperature-controlled laboratories.

An indirect application of the strain gauge forms the basis of an instrument used by Bastogne (de Meester *et al.* 1966) (Figure 8.8) to determine thicknesses of very narrow gaps, such as those found between fuel-element plates in nuclear reactors. The instrument consists of a curved metal blade with three small ruby wheels on one side and one at the centre of the opposite side. The strain gauge is attached to the outside of the blade. It indicates variations in curvature of the blade resulting from thickness changes of the gap.

## 8.4 ELECTRIFIED PARTICLE TESTING

The electrified particle method is used for testing dielectric materials; it is analogous to the magnetic particle method of testing ferromagnetic objects (Chapter 3) but it measures electrified particle

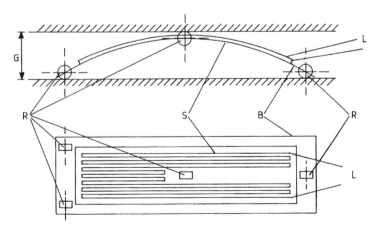

**Figure 8.8** Bastogne's strain gauge for measuring gap thicknesses between fuel-element plates in nuclear reactors. (Reprinted from de Meester *et al.*, 1966, by permission of the American Society for Nondestructive Testing)

leakage instead of magnetic flux leakage. The method has been used successfully for detecting surface cracks in glass, plastic and porcelain objects and dentures. It is often more effective than the dye penetrant technique and magnifications of the order of 30 000 times are possible. It can also be used for testing dielectric layers on metal backings such as terracotta-glazed metals and glass–metal seals. Because the method has rather specialized applications, it is treated here only briefly; further details are obtainable in McMaster (1963c).

The electrified particle method was devised by de Forest and Staats in 1945 originally to detect cracks in glass bottles. Further developments have been made by the Magnaflux Corporation, who market the technique under the tradename Statiflux. The method can be applied to dielectric materials which are either metal-backed or unbacked. In both cases, positively charged minute particles of calcium carbonate are applied from an aerosol spray to the surface of the dielectric. When a dielectric layer is backed by a metal, free negative electric charges in the metal are attracted by the sprayed positive charges and appear at the boundary surface (Figure 8.9) thus inducing localized local electric fields. When a crack appears on the surface, the thickness of the dielectric layer is reduced by an amount equal to the crack depth and the intensity of the local electric field is increased. The free negative charges on the metal surface tend to accumulate at the base of the crack, thus attracting an increased density of the positively charged particles to the upper surface of the dielectric, with the result that a magnified outline of the defect is observed.

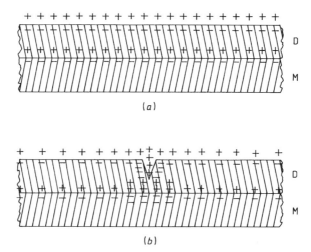

**Figure 8.9** Charge distribution on dielectric (D) and metal (M) surfaces with electrified particle testing: (a) defect-free sample, (b) sample with surface crack on dielectric layer.

With an unbacked dielectric material, the free charges are introduced by wetting the object under test with an electrically conducting liquid such as tap water. This liquid finds its way into any cracks which may be present and remains in them after the wetted surface has been allowed to dry. On spraying the surface with the positively charged calcium carbonate particles, the free charges in the liquid remaining in any cracks have the same effect on the sprayed particles as those on a backing metal surface. As before, the particle density is greatest in the vicinity of a crack and produces a magnified image of the crack.

## 8.5 DIRECT MEASUREMENT OF RESISTANCE OR CAPACITANCE

### 8.5.1 Resistance measurements

It is often necessary to measure the resistance of a homogeneous electrical conductor having a regular shape. Equation (2.1) shows that its resistance depends on its electrical resistivity (or conductivity) and dimensions. Measurements of resistance using a battery provide a simple and inexpensive means of evaluating any one of these quantities, providing the others are known. These measurements are often made for calibrating standard test samples but they can be used for determining the degree of residual stress or impurity for samples having known dimensions, or simply for measuring dimensions. It should be remembered that the resistance varies with temperature, which should be noted and kept constant during any measurements. The test sample is clamped at both ends to provide electrical contacts but electric current paths other than through the sample should be avoided, e.g. by mounting the clamp supports on a base made from an insulating material. Provided the resistance of the sample is greater than $1\,\Omega$, a relatively simple Wheatstone bridge (section 2.6.2) can be used to measure it.

Every possible care should be taken to reduce the contact resistances at the terminals but they are unlikely to exceed a small fraction of an ohm. To eliminate the effects of thermal EMFs, the polarity of the battery should be reversed, using a suitable switch, and the measurements repeated.

The Wheatstone bridge method is not suitable for samples having resistances of considerably less than $1\,\Omega$; it is not uncommon for the resistance of the sample to be less than the resistance of the electrical contacts. This difficulty can be overcome by using the Kelvin double bridge (Figure 8.10) which is energized by a direct EMF through a reversing switch so as to eliminate any contact thermal EMFs. The resistances $R_1$, $R_2$, $R_3$ and $R_4$ have values of the order of several ohms;

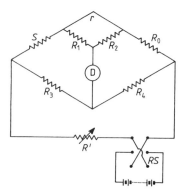

**Figure 8.10** Circuit for Kelvin double bridge.

$S$ is the test sample and $R_0$ is a standard sample having a low resistance of the same order of magnitude as $S$; $r$ represents the contact resistance between $S$ and $R_0$; $R_1$ and $R_3$ are varied in turn until there is zero potential difference across the detector, e.g. a galvanometer. It can be shown (e.g. Page and Adams, 1969) that

$$S/R_0 = R_1/R_2 = R_3/R_4 \tag{8.4}$$

The variable $r$ is eliminated during derivation. The method can be used to measure resistances as low as $1\,\mathrm{m}\Omega$.

### 8.5.2 Capacitance measurements

A simple method of determining the properties of a dielectric material in the form of a slab is to sandwich it between two parallel metal plates to form a capacitor (section 2.3.2) and to measure its electrical impedance. If the electrical impedance is purely capacitive, i.e. when the electrical permittivity $\varepsilon$ is entirely real, the value of the capacitance is $\varepsilon A/d$ (equation (2.11)).

When $\varepsilon$ is measured with the use of a capacitor, it is not always possible or convenient for the sample of dielectric material to fill completely the space between the plates. If a slab-shaped sample of thickness $t$ and cross-sectional area $A'$ is inserted between parallel metal plates each of area $A$ with a separation distance $d$ (Figure 8.11), we have effectively a capacitor $C_1$ in parallel with capacitors $C_2$ and $C_3$ in series, for which

$$C_1 = \varepsilon_0(A - A')/d \qquad C_2 = \varepsilon_0 A'/(d-t) \qquad C_3 = \varepsilon A'/t$$

Consider two capacitances $C_a$ and $C_b$. When they are connected in series, the total capacitance $C$ is given by

$$1/C = 1/C_a + 1/C_b$$

# Direct measurement of resistance or capacitance

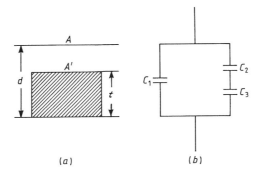

**Figure 8.11** Rectangular dielectric block between two parallel conducting plates: (a) schematic and (b) equivalent circuit.

and when in parallel

$$C = C_a + C_b$$

It can thus be shown that

$$C = \varepsilon\varepsilon_0 A'/[\varepsilon(d-t) + \varepsilon_0 t] + \varepsilon_0(A - A')/d \qquad (8.5)$$

In general, $\varepsilon$ has two components in quadrature, $\varepsilon'$ and $\varepsilon''$, such that

$$\varepsilon = \varepsilon' - j\varepsilon''$$

or dividing by $\varepsilon_0$, we get

$$\varepsilon_r = \varepsilon'_r - j\varepsilon''_r$$

Reverting, for simplicity, to the example of the dielectric block completely filling the space between the plates of the capacitor, we have

$$C = \varepsilon A/d = \varepsilon' A/d - j\varepsilon'' A/d \qquad (8.6)$$

The relationship between the potential difference $V$ across it and the current $I$ flowing through it is thus

$$I = j\omega CV = j(\omega VA/d)(\varepsilon' - j\varepsilon'')$$

Thus impedance $Z$ is given by

$$Z = R' - j/\omega C' \qquad (8.7a)$$

where $R'$ and $C'$ are the measured values of a series resistance and capacitance, respectively. It can easily be seen that

$$R' = (1/\omega C)\tan\delta/(1 + \tan^2\delta) \qquad (8.7b)$$

and

$$C' = C(1 + \tan^2\delta) \qquad (8.7c)$$

where $\tan \delta = \varepsilon''/\varepsilon'$ which is the dielectric loss factor. Here $\delta$ should not be confused with the same symbol representing logarithmic decrement (section 2.5.2) and penetration depth (section 2.8.2). In general $\tan \delta$ is a very small quantity, typically of the order of $10^{-3}$ or $10^{-4}$ (Table 2.2) and the term $\tan^2 \delta$ can usually be neglected, hence

$$C' \approx C \approx A\varepsilon'/d \qquad (8.7d)$$

and

$$R' \approx (1/\omega C) \tan \delta \qquad (8.7e)$$

One device for measuring the impedance components of a capacitor is the capacitance bridge; in its basic form it has the conditions for balance given by equation (2.57a). Consider a sheet of dielectric material in which $\varepsilon'_r$ (i.e. $\varepsilon'/\varepsilon_0$) and $\tan \delta$ are equal to 6 and 0.005, respectively. Equations (8.7d) and (8.7e) show that for a sheet thickness of 10 mm, which completely fills the space between the two capacitor plates of surface area 100 cm$^2$ and gives $C = 53$ pF, the values of $1/\omega C$ and $R'$ at a frequency of 1000 Hz are equal to 3 M$\Omega$ and 15 k$\Omega$, respectively. Taking the usual precautions to provide shielding and to prevent the effects of stray capacitances (e.g. Hague, 1934), values of $\varepsilon'_r$ and $\tan \delta$ can be obtained with a standard commercial bridge to four or more decimal places, depending on the quality of the instrument. More sophisticated bridges known as impedance analysers are available, a typical example of which is designed to operate at frequencies ranging from 5 kHz to 13 MHz and to measure capacitances of between 0.0001 pF and 100 mF and $\mathcal{Q}$ factors of between 0.1 and 1999.9. When the thickness of the sample is greater than about 10% of its smaller lateral dimension, an allowance should be made for spreading out of the electric field at the edges.

To determine the thicknesses of samples having constant values of electrical permittivity, it is only necessary to measure the capacitance. Variations of thickness can be evaluated from the output indication of the detector, previously calibrated using samples having known thicknesses.

A simpler method of capacitance testing is to connect the capacitor $C$ in series with a coil, inductance $L$ and resistance $r$, and an alternating EMF to form an $LRC$ circuit (sections 2.5.3 and 2.5.4). The total series resistance $R$ is equal to the sum of $r$ and $R'$. The frequency of the source is varied to provide a response curve (Figure 2.9) which indicates the resonance angular frequency $\omega_0 = 1/(LC)^{1/2}$ (equation 2.46a) and $\mathcal{Q} = 1/\omega_0 CR$ (equation 2.47). This is achieved using a spectrum analyser, which allows the frequency to be swept at a rapid rate, e.g. at 1000 Hz or more over the required range, and thus to provide a stationary curve on the screen of a cathode-ray oscilloscope. In practice

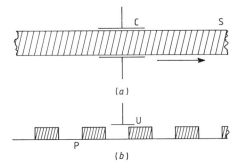

**Figure 8.12** Capacitance measurement of dielectric samples on a conveyor belt: (a) a single sheet (S) in motion and (b) an array of samples with moving upper plate (U). Other symbols: P = lower plate of the capacitor, C = conducting base. The arrow indicates the direction of motion.

the value of $r$ is negligible compared with $R$, often less than 0.1%, so $R$ can be taken as effectively equal to $R'$.

Although the degree of precision attainable may not be quite as high as with one of the more sophisticated AC bridges, it is adequate for most everyday NDT applications, including thickness gauging and observing variations in the components of $\varepsilon$ in order to detect and measure structural changes. Compared with the AC bridge, the circuit is simpler, the layout more compact and the need to avoid parasitic currents is considerably diminished. It has the further advantage of flexible response time, controlled simply by the value of the repetition frequency of the sweep. This enables measurements to be made on fast-moving conveyor systems and also provides a rapid means of data storage for computer processing. Measurements can be made on either single larger sheets or arrays of equally spaced smaller samples of dielectric materials, which can easily be conveyed on a belt between a pair of parallel plates (Figure 8.12a). Alternatively, if the dielectric rests on a metal surface, the top surface of the sample can be scanned with a conducting plate which forms the upper electrode of a capacitor (Figure 8.12b). The value of the capacitance is determined by the area $A$ of the plate surface.

Su *et al.* (1994) have reported measurements of the two components of relatively permittivity for Portland cement concrete based on the principles discussed above. The ranging frequencies were 1–40 MHz. The samples were clamped between two parallel square steel plates, side 350 mm, which could be separated from each other by 50–130 mm using a rotating screw thread.

# Appendices

## APPENDIX A

### NOTES ON UNITS

Système International (SI) units, used throughout this book, are based on the units of mass, length and time: the metre (m), the kilogram (kg) and the second (s). Multiples and submultiples of these units are related to unity by integral powers of $10^3$ using the following prefixes:

$$\begin{aligned}
\text{pico (p)} &= 10^{-12} \\
\text{nano (n)} &= 10^{-9} \\
\text{micro (µ)} &= 10^{-6} \\
\text{milli (m)} &= 10^{-3} \\
\text{kilo (k)} &= 10^{3} \\
\text{mega (M)} &= 10^{6} \\
\text{giga (G)} &= 10^{9} \\
\text{tera (T)} &= 10^{12}
\end{aligned}$$

Here are some examples:

$$\begin{aligned}
\text{1 nm (nanometre)} &= 10^{-9} \text{ m} \\
\text{1 µs (microsecond)} &= 10^{-3} \text{ s} \\
\text{1 GHz (gigahertz)} &= 10^{9} \text{ Hz}
\end{aligned}$$

Units and symbols, as used in the text, are given in Table A.1.

**Table A.1** Units of quantities used in the text.

| Quantity | Symbol[a] | Unit |
|---|---|---|
| *Mechanical units* | | |
| Mass | $m$ | kg (kilogram) |
| Length, displacement | $l, \boldsymbol{l}$ | m (metre) |
| Radius | $r, \boldsymbol{r}$ | m |
| Time | $t$ | s (second) |
| Area | $A, \boldsymbol{S}$ | m$^2$ |
| Density | $\rho$ | kg m$^{-3}$ |
| Speed, velocity | $v, \boldsymbol{v}$ | m s$^{-1}$ |
| Acceleration | $a, \boldsymbol{a}$ | m s$^{-2}$ |
| Frequency | $f$ | Hz (hertz = s$^{-1}$) |
| Angular frequency | $\omega$ | s$^{-1}$ |
| Force | $\boldsymbol{F}$ | N (newton = kg m s$^{-2}$) |
| Torque, couple | $\boldsymbol{T}$ | N m |
| Work, energy | $E$ | J (joule = N m) |
| Power | $P$ | W (watt = J s$^{-1}$) |
| Pressure | $p$ | Pa (pascal = N m$^{-2}$) |
| *Electrical and magnetic units* | | |
| Potential difference, EMF | $V, \boldsymbol{V}$ | V (volt) |
| Current | $I$ | A (ampere) |
| Charge | $Q$ | C (coulomb = A s) |
| Current density | $\boldsymbol{J}$ | A m$^{-2}$ (per unit area) |
| Charge density | $\rho$ | C m$^{-3}$ (per unit volume) |
| Impedance | $Z$ | Ω (ohm) |

| Quantity | Symbol[a] | Unit |
|---|---|---|
| *Electrical and magnetic units* | | |
| Resistance | $R$ | Ω |
| Admittance (1/Z) | $Y$ | S (siemens = Ω$^{-1}$) |
| Conductance (1/R) | $G$ | S |
| Resistivity | $\rho$ | Ω m |
| Conductivity | $\sigma$ | S m$^{-1}$ |
| Inductance (self) | $L$ | H (henry) |
| Inductance (mutual) | $M$ | H |
| Capacitance | $C$ | F (farad) |
| Electric field strength | $\boldsymbol{E}$ | V m$^{-1}$ |
| Magnetic field strength[b] | $\boldsymbol{H}$ | A m$^{-1}$ |
| Intrinsic impedance ($\boldsymbol{E}/\boldsymbol{H}$) | $Z_0$ | Ω |
| Magnetic permeability[c] | $\mu$ | H m$^{-1}$ |
| Electrical permittivity[c] | $\varepsilon$ | F m$^{-1}$ |
| Magnetic flux | $\Phi$ | Wb (weber = V s) |
| Electric flux | $\psi$ | C |
| Magnetic flux density (induction)[b] | $\boldsymbol{B}$ | T (tesla = Wb m$^{-2}$) |
| Magnetic vector potential | $\boldsymbol{A}$ | W m$^{-1}$ |
| Dielectric displacement | $\boldsymbol{D}$ | C m$^{-2}$ |
| Magnetomotive force (MMF) | $f_m$ | C m$^{-1}$ |
| Reluctance | $R_m$ | A Wb$^{-1}$ |
| Attenuation coefficient (distance) | $\alpha$ | m$^{-1}$ |
| Attenuation coefficient (time) | $\alpha'$ | s$^{-1}$ |

[a] Most of the symbols used in this book for quantities are those recommended by the International Electrotechnical Commission (IEC) and the British Standards Institution (BSI) but it often happens, due to the shortcomings of the Roman alphabet, that duplication arises. Hence, to avoid confusion, modifications may be necessary. Where appropriate, some of these symbols are expressed in vector notation, e.g. $\boldsymbol{B}$ ($V$ takes the scalar form in electromagnetic theory but becomes a vector quantity when applied to AC circuits).

[b] The traditional CGS units are still used in some quarters for $\boldsymbol{H}$ and $\boldsymbol{B}$, i.e. the oersted (Oe) (= 1000/4π A m$^{-1}$) and the gauss (G) (= 10$^4$ T).

[c] Values of $\mu$ and $\varepsilon$ for free space are 4π × 10$^{-7}$ H m$^{-1}$ and 1/(36π × 10$^9$) F m$^{-1}$, respectively.

# APPENDIX B

## STANDARDS

British Standards (BS) and American Standards (ASTM), relating to magnetic flux leakage and eddy current methods of testing are given below. National standards are currently being harmonized across the whole of Europe, and British Standards are no exception. Harmonized standards will eventually be identified by the initials BS EN; for example, BS 5411 has been revised and is now known as BS EN 2360. Harmonization is unlikely to be completed before 2001. The year of updating a British Standard is given in brackets. ASTM standards are published annually and updated when necessary.

### B.1 FLUX LEAKAGE METHODS (INCLUDING MAGNETIC PARTICLE INSPECTION)

*British Standards (BS)*

BS 6072:1981 (1986) Magnetic particle flaw detection
BS 4489:1984 Black light measurement
BS 5044:1973 (1987) Contrast aid paints
BS 5138:1974 (1988) Forged and stamped crankshafts
BS 3683 (part 2):1985 Glossary
BS 4069:1982 Inks and powders

*American Society for Testing and Materials (ASTM)*

ASTM E 709 Magnetic particle inspection practice
ASTM E 125 Indications in ferrous castings
ASTM E 1316 Definition of terms
ASTM E 570 Flux leakage examination of ferromagnetic steel tubular products

*Appendix B*

## B.2 EDDY CURRENT METHODS

*British Standards (BS)*

BS 3683 (part 5):1965 (1989) Eddy current flaw detection glossary
BS 3889 (part 2A):1986 (1991) Automatic eddy current testing of wrought steel tubes
BS 3889 (part 2B):1966 (1987) Eddy current testing of non-ferrous tubes
BS 5411 (part 3):1984 Eddy current methods for measurement of coating thickness of non-conductive coatings on non-magnetic base material. Withdrawn: now known as BS EN 2360 (1995).

*American Society for Testing and Materials (ASTM)*

ASTM A 450/A450M General requirements for carbon, ferritic alloys and austenitic alloy steel tubes
ASTM B 244 Method for measurement of thickness of anodic coatings of aluminum and other nonconductive coatings on nonmagnetic base materials with eddy current instruments
ASTM B 659 Recommended practice for measurement of thickness of metallic coatings on nonmetallic substrates
ASTM E 215 Standardising equipment for electromagnetic testing of seamless aluminium alloy tube
ASTM E 243 Electromagnetic (eddy current) testing of seamless copper and copper alloy tubes
ASTM E 309 Eddy current examination of steel tubular products using magnetic saturation
ASTM E 376 Measuring coating thickness by magnetic field or eddy current (electromagnetic) test methods
ASTM E 426 Electromagnetic (eddy current) testing of seamless and welded tubular products austenitic stainless steel and similar alloys
ASTM E 566 Electromagnetic (eddy current) sorting of ferrous metals
ASTM E 571 Electromagnetic (eddy current) examination of nickel and nickel alloy tubular products
ASTM E 690 In-situ electromagnetic (eddy current) examination of non-magnetic heat-exchanger tubes
ASTM E 703 Electromagnetic (eddy current) sorting of nonferrous metals
ASTM E 1004 Electromagnetic (eddy current) measurements of electrical conductivity
ASTM E 1033 Electromagnetic (eddy current) examination of type F continuously welded (CW) ferromagnetic pipe and tubing above the Curie temperature

ASTM E 1316 Definition of terms relating to electromagnetic testing
ASTM G 46 Recommended practice for examination and evaluation of pitting corrosion

# APPENDIX C

## BESSEL FUNCTIONS

### C.1 GENERAL CONSIDERATIONS

A general account of Bessel functions applicable to engineering problems has been given by McLachlan (1934). This section provides a summary of their properties as applied to electrical methods of testing, with particular reference to eddy currents.

Here is a modified form of Bessel's equation:

$$d^2Z/dr^2 + (1/r)\,dZ/dr + k^2Z = 0 \tag{C.1}$$

where $k$ is a constant. One solution of this equation is

$$Z = A_1 J_0(kr) + B_1 Y_0(kr) \tag{C.2}$$

where $A_1$ and $B_1$ are constants, and $J_0(kr)$ and $Y_0(kr)$ represent infinite series and are called zeroth-order Bessel functions of the first and second kind, respectively, and take the forms of infinite convergent series. The series for $J_0(kr)$ is given in equation (C.6a).

In eddy current analyses the variable $k$ is often complex, as defined by $k^2 = -j|k|^2$ (equation 4.4). Taking the positive root for which $k = j^{3/2}|k|$, where $j = (-1)^{1/2}$, equation (C.2) becomes

$$Z = A_1 J_0(j^{3/2}|kr|) + B_1 K_0(j^{3/2}|kr|) \tag{C.2*}$$

The use of $Y_0(kr)$ is not generally desirable for applications of interest here because it equals infinity when $r$ becomes infinite. The following solution has more practical uses:

$$Z = A_1 J_0(j^{3/2}|kr|) + B_1 K_0(j^{1/2}|kr|) \tag{C.3}$$

where $A_1$ and $B_1$ are again constants but not necessarily having the same values as those in equation (C.2). $K_0(kr)$ is a modified zeroth-order Hankel function which has the advantage of decreasing asymptotically to zero as $r$ approaches infinity. For the induction of eddy currents in cylindrical rods (section 4.2), a complete solution to equation (C.3) is given by an equation of the form

$$Z = A_1 J_0(j^{3/2}|kr|) \tag{C.3*}$$

Also of interest is a first-order Bessel equation of the following form:

$$\partial^2 Z/\partial r^2 + (1/r)\partial Z/\partial r + Z(k^2 - 1/r^2) = 0 \quad \text{(C.4)}$$

where $k$ is a constant. A solution of interest is

$$Z = A_2 J_1(kr) + B_2 Y_1(kr) \quad \text{(C.5)}$$

where $A_2$ and $B_2$ are constants and $J_1$ and $Y_1$ are first-order Besssel functions of the first and second kind, respectively.

## C.2  THE J FUNCTIONS

The series represented by the functions $J_0(kr)$ and $J_1(kr)$ are given by the following expressions:

$$J_0(kr) = \sum_{n=0}^{\infty} (-1)^n (kr)^{2n} / 2^{2n} (n!)^2 \quad \text{(C.6a)}$$

and

$$J_1(kr) = \sum_{n=0}^{\infty} (-1)^n (kr)^{(2n+1)} / 2^{(2n+1)} n!(n+1)! \quad \text{(C.6b)}$$

from which it can be seen that

$$\int_0^r r J_0(kr) dr = r^2 J_1(kr)/kr \quad \text{(C.7)}$$

and

$$J_1(kr) = -d/dr[J_0(kr)] = -J_0'(kr) \quad \text{(C.8)}$$

When $k$ has the complex value $j^{3/2}|k|$, expanding the zeroth-order Bessel function into its real and imaginary parts gives

$$J_0(kr) = J_0(j^{3/2}|kr|) = \text{ber}|kr| + j\,\text{bei}|kr| \quad \text{(C.9a)}$$

and

$$J_0'(kr) = J_0'(j^{3/2}|kr|) = j^{-3/2}(\text{ber}'|kr| + j\,\text{bei}'|kr|) \quad \text{(C.9b)}$$

where ber and bei respectively indicate the real and imaginary components of $J_0$ and ber' and bei' those of $J_0'$; remember the multiplying factor $j^{-3/2}$ for $J_0'$. These components can easily be derived from equations (C.6) and (C.8), and the corresponding series are

$$\text{ber}\,|kr| = \sum_{n=0}^{\infty} (-1)^n |kr|^{2r} \cos(3\pi n/2)/2^{2n}(n!)^2 \tag{C.10}$$

$$\text{bei}\,|kr| = \sum_{n=0}^{\infty} (-1)^n |kr|^{2r} \sin(3\pi n/2)/2^{2n}(n!)^2 \tag{C.11}$$

$$\text{ber}'\,|kr| = \sum_{n=0}^{\infty} (-1)^n n |kr|^{2n-1} \cos(3\pi n/2)/2^{2n-1}(n!)^2 \tag{C.12}$$

$$\text{bei}'\,|kr| = \sum_{n=0}^{\infty} (-1)^n n |kr|^{2n-1} \sin(3\pi n/2)/2^{2n-1}(n!)^2 \tag{C.13}$$

For higher values of $|kr|$ we can use the approximations given below. McLachlan (1934) suggests their use for $|kr|$ greater than 10, when accuracies to about 0.1% may be obtained; however, with the aid of a personal computer, the author has been able to raise this limit to 20 without encountering any difficulties, improving the accuracy to better than 1 part in $10^4$

$$\text{ber}\,|kr| \approx A[\sin(B + 3\pi/8) + (1/8|kr|)\sin(B + \pi/8)] \tag{C.14}$$

$$\text{bei}\,|kr| \approx A[\sin(B - \pi/8) + (1/8|kr|)\sin(B - 3\pi/8)] \tag{C.15}$$

$$\text{ber}'\,|kr| \approx A[\sin(B + 5\pi/8) - (3/8|kr|)\sin(B + 3\pi/8)] \tag{C.16}$$

$$\text{bei}'\,|kr| \approx A[\sin(B + \pi/8) - (3/8|kr|)\sin(B - \pi/8)] \tag{C.17}$$

where $A = 0.3989\,\exp(|kr|/2^{1/2})/|kr|^{1/2}$ and $B = 0.7071\,|kr|$ in radians.

## C.3 THE $K_0$ FUNCTION

The function $K_0(j^{1/2}|kr|)$ is more conveniently expressed in terms of its real and imaginary components, i.e.

$$K_0(j^{1/2}|kr|) = \text{ker}\,|kr| + j\,\text{kei}\,|kr| \tag{C.18}$$

and

$$K_0'(j^{1/2}|kr|) = j^{-1/2}(\text{ker}'\,|kr| + j\,\text{kei}'\,|kr|) \tag{C.19}$$

where $K_0'(j^{1/2}|kr|) = (d/dr)K_0(j^{1/2}|kr|)$. The ker, kei, ker′, and kei′ functions are defined in the same way as the ber, bei, ber′ and bei′ functions. We have

$$\ker|kr| = (\log_e 2 - g - \log_e |kr|) \operatorname{ber} |kr| + (\pi/4) \operatorname{bei} |kr|$$
$$- (|kr|/2)^4(1+\tfrac{1}{2})/(2!)^2 + (|kr|/2)^8(1+\tfrac{1}{2}+\tfrac{1}{3}+\tfrac{1}{4})/(4!)^2 - \cdots$$
(C.20)

$$\ker|kr| = (\log_e 2 - g - \log_e |kr|) \operatorname{bei} |kr| - (\pi/4) \operatorname{bei} |kr| + (|kr|/2)^2$$
$$- (|kr|/2)^6(1+\tfrac{1}{2}+\tfrac{1}{3})/(3!)^2 + \cdots \qquad (C.21)$$

where $g$ is Euler's constant $= 0.577\,215\,665$.

Values of ker′ $|kr|$ and kei′ $|kr|$ can be obtained by differentiating, term by term, the series for ker $|kr|$ and kei $|kr|$, respectively.

For values of $|kr|$ greater than 10, the following approximations can be used:

$$\ker|kr| \approx C[\sin(B + 5\pi/8) + (\tfrac{1}{8}|kr|)\sin(D - \pi/8)] \qquad (C.22)$$

$$\ker|kr| \approx C[-\sin(B + \pi/8) + (\tfrac{1}{8}|kr|)\sin(D - 3\pi/8)] \qquad (C.23)$$

$$\ker'|kr| \approx -C[\sin(B + 3\pi/8) + (\tfrac{3}{8}|kr|)\sin(D - 5\pi/8)] \qquad (C.24)$$

$$\ker'|kr| \approx C[\sin(B - \pi/8) + (\tfrac{3}{8}|kr|)\sin(D + \pi/8)] \qquad (C.25)$$

where $C = 1.2533 \exp(-|kr|/2^{1/2})/|kr|^{1/2}$ and $D = 0.7071|kr|$ in radians.

# APPENDIX D

## BASIC PROGRAMS FOR PREDICTING IMPEDANCE COMPONENTS*

### D.1 COILS ENCIRCLING ELECTRICALLY CONDUCTING CYLINDRICAL RODS**

```
100 PRINT TAB(4) "PREDICTIONS OF NORMALISED IMPEDANCES FOR COILS"
110 PRINT TAB(17)           "ENCIRCLING METAL RODS"
120 PRINT TAB(4) "================================================
130 PRINT
140 PRINT "Option 1: input values absolute"
150 PRINT "Option 2: input values normalised"
160 INPUT "Type option number............",i
170 PRINT
180 IF i=2 THEN 290
190 INPUT "Elec. cond. in MS/m = ", k
200 INPUT "Rel. mag. perm. = ", m
210 INPUT "Coil radius in mm = ", j
220 INPUT "Radius of rod in mm = ", g
230 INPUT "Frequency in kHz = ", n
240 h=8*n*m*k*g*g*PI*PI/10^4
250 PRINT "Normalised frequency = ";USING "#####.##";h
260 v = (g/j)^2*100
270 PRINT "Percentage fill-factor = ";USING "##.##";v
280 GOTO 320

290 INPUT "Rel.mag. perm. = ",m
300 INPUT "Normalised frequency = ", h
310 INPUT "Percentage fill factor = ", v
320 z=h^.5
330 IF z>20 THEN 590
340 a=1
350 p=1
360 FOR r=2 TO 60 STEP 2
370 p=-(z/2)^4*p/r^2/(r-1)^2
380 a=a+p
390 NEXT r

400 b=(z/2)^2
410 q=(z/2)^2
420 FOR r=3 TO 59 STEP 2
430 q=-(z/2)^4*q/r^2/(r-1)^2
440 b=b+q
450 NEXT r

460 c=-(z/2)^3/2
470 s=-(z/2)^3/2
480 FOR r=4 TO 60 STEP 2
490 s=-(z/2)^4*s/r/(r-2)/(r-1)^2
500 c=c+s
510 NEXT r
```

*Some computer software does not recognize the symbol PI, so the number 3.141 89 should be substituted where appropriate.
**After Blitz (1989). Reproduced by permission of Butterworth-Heinemann Ltd ©.

```
520 d=z/2
530 t=z/2
540 FOR r=3 TO 59 STEP 2
550 t=-(z/2)^4*t/r/(r-2)/(r-1)^2
560 d=d+t
570 NEXT r
580 GOTO 630

590 a=SIN(PI*z*9/40+3*PI/8)+1/8/z*SIN(PI*z*9/40+PI/8)
600 b=SIN(PI*z*9/40-PI/8)+1/8/z*SIN(PI*z*9/40-3*PI/8)
610 c=SIN(PI*z*9/40+5*PI/8)-3/8/z*SIN(PI*z*9/40+3*PI/8)
620 d=SIN(PI*z*9/40+PI/8)-3/8/z*SIN(PI*z*9/40-PI/8)

630 e=(2/z)*(a*c+b*d)/(a^2+b^2)
640 f=(2/z)*(a*d-b*c)/(a^2+b^2)
650 x=v/100*m*e
660 y=v/100*m*f+1-v/100
670 PRINT "wL/wLo = ";USING "#####.####";y
680 PRINT "R/wLo = ";USING "#####.####";x
690 END
```

This program offers the choice of inserting either absolute or normalized values of the quantities required for calculating the normalized values of impedance of an infinitely long coil encircling an infinitely long rod. The variable $z$ which represents $\beta$ (equation 4.14) is defined in line 320 and, for values of less than 20 the ber, bei, ber' and bei' functions are evaluated in lines 340 to 390, 400 to 450, 460 to 510 and 520 to 570, respectively. Values of $z$ greater than 20 are covered by the explicit equations contained in lines 590 to 620. Lines 630 and 640 contain the expressions for $\mu_R$ and $\bar{\mu}_I$ (equations 4.16) and lines 650 and 660 represent equations (4.22). The results are correct to four decimal places, but by reducing the maximum value of $r$ to 30 and by using lines 590 to 620 for values of $z$ greater than 10, the accuracy is reduced to three decimal places.

## D.2 COILS ENCIRCLING ELECTRICALLY CONDUCTING CYLINDRICAL TUBES

```
100 PRINT "PREDICTION OF COMPONENTS OF NORMALISED IMPEDANCE OF COILS"
110 PRINT TAB(12) "ENCIRCLING DEFECT-FREE METAL TUBES"
120 PRINT  "================================================================"
130 DIM a(6),b(6),c(6),d(6),e(12),g(6),h(3),i(12),l(3),m(3),n(4)
140 DIM u(12),v(2),w(2),x(12),y(2)
150 PRINT
160 PRINT "Option 1: input values absolute"
170 PRINT "Option 2: input values normalised"
180 INPUT "Type option number......... ",o
190 IF o=2 THEN 330

200 INPUT "Elec. cond. in MS/m = ",m(3)
210 INPUT "Rel. mag. perm. = ",m(1)
220 INPUT "Coil radius in mm = ",g(5)
230 INPUT "External radius of tube in mm = ",g(6)
240 INPUT "Internal radius of tube in mm = ",g(4)
250 INPUT "Frequency in kHz = ",y(2)
260 y(1)=8*y(2)*m(1)*(g(6)*PI)^2/10^4
270 PRINT "Normalised frequency = ";USING "####.##";y(1)
280 g(1)=g(4)/g(6)
290 PRINT "Diameter ratio = ";g(1)
300 m(2)=(g(6)/g(5))^2*100
310 PRINT "Percentage fill-factor = ";USING "##.##";m(2)
320 GOTO 370

330 INPUT "Normalised frequency = ",y(1)
340 INPUT "Rel. mag. perm. = ",m(1)
350 INPUT "Diameter ratio = ",g(1)
360 INPUT "Percentage fill-factor = ",m(2)
370 g(2)=1

380 z=y(1)^.5
390 GOSUB 860

400 a(2)=a(1)
410 b(2)=b(1)
420 c(2)=c(1)
430 d(2)=d(1)
440 a(5)=a(4)
450 b(5)=b(4)
460 c(5)=c(4)
470 d(5)=d(4)
480 z=y(1)^.5*g(1)
490 GOSUB 860
```

```
500 a(3)=a(1)
510 b(3)=b(1)
520 c(3)=c(1)
530 d(3)=d(1)
540 a(6)=a(4)
550 b(6)=b(4)
560 c(6)=c(4)
570 d(6)=d(4)

580 n(1)=(y(1)^.5)*g(1)/2
590 i(1)=(n(1)*(a(6)-b(6))-c(6)-d(6))/2^.5
600 e(1)=(n(1)*(a(6)+b(6))+c(6)-d(6))/2^.5
610 i(2)=(n(1)*(b(3)-a(3))+c(3)+d(3))/2^.5
620 e(2)=(d(3)-c(3)-n(1)*(a(3)+b(3)))/2^.5
630 i(3)=i(1)*a(2)-e(1)*b(2)+i(2)*a(5)-e(2)*b(5)
640 e(3)=i(1)*b(2)+e(1)*a(2)+i(2)*b(5)+e(2)*a(5)
650 i(4)=i(1)*i(3)+e(1)*e(3)
660 e(4)=i(3)*e(1)-e(3)*i(1)
670 i(5)=i(2)*i(3)+e(2)*e(3)
680 e(5)=-i(2)*e(3)+e(2)*i(3)
690 n(2)=g(2)^2/y(1)^.5
700 n(3)=g(1)*n(2)
710 i(6)=n(2)*d(2)-n(3)*d(3)
720 e(6)=n(3)*c(3)-n(2)*c(2)
730 i(7)=n(3)*d(6)-n(2)*d(5)
740 e(7)=n(2)*c(5)-n(3)*c(6)
750 n(4)=i(3)^2+e(3)^2
760 i(8)=(i(4)*i(6)-e(4)*e(6)+i(5)*i(7)-e(5)*e(7))/n(4)
770 e(8)=(i(4)*e(6)+e(4)*i(6)+e(5)*i(7)+i(5)*e(7))/n(4)
780 i(9)=(i(4)*a(3)-e(4)*b(3)+i(5)*a(6)-e(5)*b(6))/n(4)
790 e(9)=(i(4)*b(3)+e(4)*a(3)+e(5)*a(6)+i(5)*b(6))/n(4)
800 i(11)=(i(9)^2+e(9)^2)^.5
810 i(12)=m(2)/100*(2*m(1)*i(8)/g(2)^2+g(1)^2*i(9)-1)+1
820 e(12)=-m(2)/100*(2*m(1)*e(8)/g(2)^2+g(1)^2*e(9))
830 PRINT "wL/wLo = ";USING "####.####"; i(12)
840 PRINT "R/wLo  = ";USING "####.####"; e(12)
850 END

860 IF z>10 THEN 1420
870 a(1)=1
880 p=1
890 FOR j=2 TO 60 STEP 2
900 p=-(z/2)^4*p/j^2/(j-1)^2
910 a(1)=a(1)+p
920 NEXT j
```

```
930 b(1)=(z/2)^2
940 q=(z/2)^2
950 FOR j=3 TO 59 STEP 2
960 q=-(z/2)^4*q/j^2/(j-1)^2
970 b(1)=b(1)+q
980 NEXT j

990 c(1)=-(z/2)^3/2
1000 r=-(z/2)^3/2
1010 FOR j=4 TO 60 STEP 2
1020 r=-(z/2)^4*r/j/(j-2)/(j-1)^2
1030 c(1)=c(1)+r
1040 NEXT j

1050 d(1)=z/2
1060 t=z/2
1070 FOR j=3 TO 59 STEP 2
1080 t=-(z/2)^4*t/j/(j-2)/(j-1)^2
1090 d(1)=d(1)+t
1100 NEXT j

1110 REM log 2 = 0.693147181
1120 REM Euler's const = 0.577215665
1130 u(1)=(.115931516-LOG(z))*a(1)+PI/4*b(1)-.375*(z/2)^4
1140 u(2)=3.61689815/10^3*(z/2)^8-4.72608025/10^6*(z/2)^12
1150 u(3)=1.67180484/10^9*(z/2)^16-2.22427561/10^13*(z/2)^20
1160 u(4)=1.35250019/10^17*(z/2)^24-4.27834083/10^22*(z/2)^28
1170 u(5)=7.72273568/10^27*(z/2)^32-8.52665261/10^32*(z/2)^36
1180 u(6)=6.07927716/10^37*(z/2)^40
1190 a(4)=u(1)+u(2)+u(3)+u(4)+u(5)+u(6)
1200 x(1)=(.115931516-LOG(z))*b(1)-PI/4*a(1)+(z/2)^2
1210 x(2)=-5.09259259/10^2*(z/2)^6+1.58564815/10^4*(z/2)^10
1220 x(3)=-1.0207456/10^7*(z/2)^14+2.14833502/10^11*(z/2)^18
1230 x(4)=-1.89522959/10^15*(z/2)^22+8.20193883/10^20*(z/2)^26
1240 x(5)=-1.9047089/10^24*(z/2)^30+2.71872272/10^29*(z/2)^34
1250 x(6)=-2.39752144/10^34*(z/2)^38+1.41395064/10^39*(z/2)^42
1260 b(4)=x(1)+x(2)+x(3)+x(4)+x(5)+x(6)
1270 u(7)=(.115931516-LOG(z))*c(1)-a(1)/z+PI/4*d(1)-1.5*z^3/2^4
1280 u(8)=3.61689815/10^3*8*z^7/2^8-4.72608025/10^6*12*z^11/2^12
1290 u(9)=1.67180484/10^9*16*z^15/2^16-2.22427561/10^13*20*z^19/2^20
1300 u(10)=1.35250019/10^17*24*z^23/2^24-4.27834083/10^22*28*z^27/2^28
1310 u(11)=7.72273568/10^27*32*z^31/2^32-8.52665261/10^32*36*z^35/2^36
1320 u(12)=6.07927716/10^37*40*z^39/2^40
1330 c(4)=u(7)+u(8)+u(9)+u(10)+u(11)+u(12)
```

## Appendix D

```
1340 x(7)=(.115931516-LOG(z))*d(1)-b(1)/z-PI/4*c(1)+2*z/4
1350 x(8)=-5.09259259/10^2*6*z^5/2^6+1.58564815/10^4*10*z^9/2^10
1360 x(9)=-1.0207456/10^7*14*z^13/2^14+2.14833502/10^11*18*z^17/2^18
1370 x(10)=-1.89522959/10^15*22*z^21/2^22+8.20133883/10^20*26*z^25/2^26
1380 x(11)=-1.9047089/10^24*30*z^29/2^30+2.71872272/10^29*34*z^33/2^34
1390 x(12)=-2.39752144/10^34*38*z^37/2^38+1.41395064/10^39*42*z^41/2^42
1400 d(4)=x(7)+x(8)+x(9)+x(10)+x(11)+x(12)
1410 GOTO 1520

1420 h(1)=40.514*PI/180*z
1430 h(2)=.3989*EXP(z/2^.5)/z^.5
1440 a(1)=h(2)*(SIN(h(1)+3*PI/8)+1/8/z*SIN(h(1)+PI/8))
1450 b(1)=h(2)*(SIN(h(1)-PI/8)+1/8/z*SIN(h(1)-3*PI/8))

1460 c(1)=h(2)*(SIN(h(1)+5*PI/8)-3/8/z*SIN(h(1)+3*PI/8))
1470 d(1)=h(2)*(SIN(h(1)+PI/8)-3/8/z*SIN(h(1)-PI/8))

1480 h(3)=1.2533*EXP(-z/2^.5)/z^.5
1490 a(4)=h(3)*(SIN(h(1)+5*PI/8)+1/8/z*SIN(h(1)-PI/8))
1500 b(4)=h(3)*(-SIN(h(1)+PI/8)+1/8/z*SIN(h(1)+3/PI/8))
1510 c(4)=-h(3)*(SIN(h(1)+3*PI/8)+3/8/z*SIN(h(1)+5*PI/8))
1520 d(4)=h(3)*(SIN(h(1)-PI/8)+3/8/z*SIN(h(1)+PI/8))
1530 RETURN
```

This program also offers the choice of inserting either absolute or normalized values of frequency and diameter ratio. In preparing the program, account had to be taken of the value of unity assigned to g(2), representing the external radius $b$ of the tube (section 4.3, penultimate paragraph). To avoid complications arising with the 'absolute values' option, the specific values for external and internal radii, i.e. g(6) and g(4), are included in lines 230 and 240. These were used solely to derive g(1) (line 280); they play no further part in the calculations.

The calculations for the complex functions are provided in a subroutine, lines 860 to 1520, which includes the derivations of the functions ber, bei, ber' and bei' as before, indicated by a(1), b(1), c(1) and d(1), and also of the functions ker, kei, ker' and kei', represented by a(4), b(4), c(4) and d(4). The nature of the series which describe these functions does not allow any simplified means of computation, except for values of $z$ greater than 10, and the series for each of these functions was expanded until overrunning occurred, a factor which restricted the upper value of $z$. For this reason, accuracy cannot be guaranteed to better than three places of decimals for higher normalized frequencies.

Lines 400 to 570 define the functions of $\alpha$ and $\beta$, the array numerals 2 and 3 corresponding to $\beta$ and 5 and 6 to $\alpha$. Lines 210 and 310 define the values of $\beta$ and $\alpha$, respectively; i(1) and e(1) are

*Appendix D*

the real and imaginary components respectively of $G$, i(2) and e(2) of $F$ and i(3) and e(3) of $D$ (equation 4.32). Lines 410 to 630 perform substitutions and rationalizations, and the values of the real and imaginary components of $\bar{B}/B_0$ and $B_i/B_0$, are given by i(8) and e(8), and i(9) and e(9), respectively.

## D.3  AIR-CORED COILS: SCANNING THE SURFACES OF ELECTRICAL CONDUCTORS*

```
100 PRINT "PREDICTION OF NORMALISED IMPEDANCES FOR AIR- CORED"
110 PRINT "COILS SCANNING METAL SURFACES"
120 PRINT "===================================================="
130 PRINT
140 PRINT "Option 1: input values absolute"
150 PRINT "Option 2: input values normalised"
160 DIM q(3),h(2)
170 INPUT "Type option number...........",p
180 IF p=2 THEN 320
190 INPUT "Elec. cond. in MS/m = ",q(1)
200 INPUT "Rel. mag. perm.=",q(3)
210 INPUT "Coil radius in mm = ",j
220 INPUT "Coil length in mm = ",h(1)
230 INPUT "Lift-off in mm = ",h(2)
240 INPUT "Frequency in kHz = ",q(2)
250 e=8*q(1)*q(2)*q(3)*(PI*j)^2/10^4
260 c=h(1)/j
270 d=h(2)/j
280 PRINT "Normalised frequency = ";USING "#####.##";e
290 PRINT "Normalised coil length = ";USING "#.##";c
300 PRINT "Normalised lift-off = ";USING "#.##";d
310 GOTO 360

320 INPUT "Normalised coil length ko = ",c
330 INPUT "Rel. mag. perm. =",q(3)
340 INPUT "Normalised lift-off k = ",d
350 INPUT "Normalised frequency fo = ",e
360 DIM s(3),u(3),i(2),m(3),w(3)
370 s(1)=0
380 s(2)=0
390 s(3)=0
400 u(1)=0
410 u(2)=0
420 u(3)=0
430 x=0
440 GOSUB 500
450 i(1)=1+w(1)/(c-8/3/PI+2*w(2))
460 i(2)=w(3)/(c-8/3/PI+2*w(2))
470 PRINT"wL/wL0 = ";USING "#.####";i(1)
480 PRINT"R/wL0 = ";USING "#.####";i(2)
490 PRINT
500 PRINT
510 END
```

*Reproduced by permission of Dr S. R. Oaten.

```
520 FOR k=1 TO 499
530 x=x+.02
540 GOSUB 730
550 IF k/2-INT(k/2)=0 THEN 600
560 s(1)=s(1)+f
570 s(2)=s(2)+v
580 s(3)=s(3)+b
590 GOTO 630

600 u(1)=u(1)+f
610 u(2)=u(2)+v
620 u(3)=u(3)+b
630 NEXT k
640 x=10
650 GOSUB 710
660 m(1)=f
670 m(2)=v
680 m(3)=b
690 w(1)=(m(1)+s(1)*4+u(1)*2)/150
700 w(2)=(m(2)+s(2)*4+u(2)*2)/150
710 w(3)=(m(3)+s(3)*4+u(3)*2)/150
720 RETURN

730 a=x/2
740 t=x/2
750 FOR r=1 TO 15
760 t=-(x/2)^2*t/r/(r+1)
770 a=a+t
780 NEXT r
790 y=(a/x)^2
800 g=SQR(x^4+e^2)
810 n=x^2+g+x*SQR(2*(g+x^2))
820 z=y/EXP(2*d*x)*(1-1/EXP(c*x))^2
830 f=z*(x^2-g)/n
840 v=y/EXP(c*x)
850 b=z*x*q(3)*SQR(2*(g-x^2))/n
860 RETURN
```

Also valid for ferrite-cored surface-scanning coils with non-ferromagnetic conductors using low amplitude excitation currents, this program accepts absolute or normalized data. The integrals are solved in a subroutine (lines 520 to 720) using Simpson's rule. The number of steps allowed here for this purpose is 500, which must be regarded as a minimum and is just about sufficient to give values correct to four decimal places. Increasing the number of steps should increase the degree of precision but will also increase the time of computation. If this is done, the necessary amendments must be made to steps 520, 530, 690, 700 and 710.

## Appendix D

The Bessel functions are calculated in another subroutine (lines 730 to 780) which is continued (lines 790 to 860) to evaluate the integrands. the expression $(\mu_r X - X_1)/(\mu_r X + X_1)$ (cf. equation 4.54) is separated into real and imaginary parts, as follows:

Real part
$$[(\mu_r X)^2 - G]/N \qquad \text{(line 830)}$$

Imaginary part
$$\mu_r X[2(G - X^2)]^{1/2}/N \qquad \text{(line 850)}$$

where
$$G = (X^4 + \beta^4)^{1/2} \qquad \text{(line 800)}$$

and
$$N = (\mu_r X)^2 + G + \mu_r X[2(G + X^2)]^{1/2} \qquad \text{(line 810)}$$

# References

Adam S F (1969) *Microwave Theory and Applications* (Englewood Cliffs, NJ: Prentice Hall).
Aldeen A and Blitz J (1979) *NDT Int.* **12** 211–16.
Aleksander I and Morton H (1990) *An Introduction to Neural Computing* (London: Chapman & Hall).
Anderson J C (1968) *Magnetism and Magnetic Materials* (London: Chapman & Hall).
Atherton D L (1983) *NDT Int.* **16** 145–49.
Atherton D L, Toal C J and Schmidt T R (1989) *Br. J. NDT* **31** 485–88.
Auld B A (1981) *Eddy Current Characterisations of Materials and Structures* ASTM STP722, eds B Birnbaum and G Free (Philadelphia, PA: American Society for Testing and Materials) pp 332–47.
Auld B A and Winslow D K (1981) *Eddy Current Characterisations of Materials and Structures* ASTM STP722, eds G Birnbaum and G Free (Philadelphia, PA: American Society for Testing and Materials) pp 348–66.
Aurin W F (1996) *Review of Progress in Nondestructive Evaluation*, vol 15A, eds D O Thompson and D E Chimento, (New York: Plenum) pp 1145–50.
Bahr A J (1981) *Eddy Current Characterisations of Materials and Structures* ASTM STP722, eds G Birnbaum and G Free (Philadelphia, PA: American Society for Testing and Materials) pp 311–31.
Bahr A J (1983) *Microwave Nondestructive Testing Methods* (New York: Gordon and Breach).
Bailey A E (1989) *Microwave Measurements*, 2nd edn (London: Peter Peregrinus).
Bareham F R (1960) *Br. J. Appl. Phys.* **11** 218–22.
Becker R, Betzhold K, Boness K D, Collins R, Holt C C and Simkin J (1986) *Br. J. NDT* **28** 286–94 and 361–70.
Bednorz J G and Müller K A (1986) *Z. Phys.* **64** 189–93.
Bell W W (1968) *Handbook of Special Functions for Scientists and Engineers* (London: Van Nostrand).
Bergander M J (1985) *Electromagnetic Methods of Nondestructive Testing*, ed W Lord (New York: Gordon and Breach) pp 21–33.
Blitz J (1971) *Ultrasonics: Methods and Applications* (London: Butterworth).
Blitz J (1989) *NDT Int.* **22** 3–6.
Blitz J and Alagoa K D (1985) *NDT Int.* **18** 269–73.
Blitz J and Simpson G (1996) *Ultrasonic Methods of Non-destructive Testing*, (London: Chapman & Hall) pp 73–76.
Blitz J, King W G and Rogers D G (1969a) *Electrical, Magnetic and Visual Methods of Testing Materials* (London: Butterworth) Chs 3, 4.
Blitz J, King W G and Rogers D G (1969b) *Electrical, Magnetic and Visual Methods of Testing Materials* (London: Butterworth) pp 12–13.
Blitz J, Oaten S R and Hajian N T (1986) *Nondestruct. Testing Commun.* **2** 189–200.
Blitz J, Williams D J A and Tilson J P (1981) *NDT Int.* **14** 119–23.

Born M and Wolf W (1970) *Principles of Optics*, 4th edn (Oxford: Pergamon).
Botsco R J and McMaster R C (1996) *Nondestructive Testing Handbook*, vol 4, 2nd edn, ed R C McMaster (Columbus, OH: American Society for Nondestructive Testing) sections 16 and 17.
Botsco R J, Cribbs R W, King R J and McMaster R C (1986) *Nondestructive Testing Handbook*, vol 4, 2nd edn, ed R C McMaster (Columbus, OH: American Society for Nondestructive Testing) section 18.
Bridge B (1987) *Nondestruct. Testing Commun.* **3** 47–55.
Brenneke H (1989) *Br. J. NDT* **31** 87–90.
Brown D J and Le Q V (1989) *Mater. Eval.* **47** 47–55.
Buckingham H and Price E M (1966) *Principles of Electrical Measurements*, 2nd edn (London: EUP).
Bungey J H and Millard S G (1995) *Nondestruct. Test. Eval.* **12** 33–51.
Chatterjee R (1988) *Advanced Microwave Engineering* (Chichester: Ellis Horwood).
Cecco V S, Van Drunen G and Sharp F L (1981) *Eddy Current Testing: Manual of Eddy Current Methods*, vol 1, report AECL-7523, (Chalk River, Ont.: Atomic Energy of Canada Limited).
Charlesworth F D W and Dover W D (1982) *Advances in Crack Length Measurement*, ed C J Beevers (Warley: EMAS) pp 253–76.
Charlton P C (1993) *Insight* **35** 433–37.
Chedister W C (1994) *Mater. Eval.* **52** 934–38.
Chen H L R, Halebe U B, Sami Z and Vasudez B (1994) *Mater. Eval.* **52** 1382–88.
Clark R and Bond L J (1989) *NDT Int.* **22** 331–38.
Clarke J (1994) *Scientific American* **271** 36–43 (August).
Clayden N L and Jackson P (1994) *Mater. Eval.* **11** 293–304.
Collins R, Dover W D and Michael D H (1985) *Research Techniques in Nondestructive Testing*, vol 8, ed R S Sharpe (London: Academic) Ch 5.
Collins R, Niemero A and Lewis A M (1990) *Advances in Underwater Inspection and Maintenance* (*Advances in Underwater Technology, Ocean Science and Off-shore Engineering*, vol 21) (London: Graham and Trotman) Ch 9.
Cooper P (1983) *Br. J. NDT* **25** 75–83.
Cottrell A H (1964) *The Mechanical Properties of Matter* (New York: Wiley).
Davis T J (1981) *Eddy Current Characterisations of Materials and Structures* ASTM STP722, eds G Birnbaum and G Free (Philadelphia, PA: American Society for Testing and Materials) pp 255–65.
Dean D S and Kerridge L A (1970) *Research Techniques in Nondestructive Testing*, ed R S Sharpe (London: Academic) Ch 13.
de los Santos (1994) *Mater. Eval.* **11** 305–11.
de Meester P, Decknock R and Verstapen G (1966) *Mater. Eval.* **24** 482–86.
Deutsch V and Vogt M (1982) *Br. J. NDT* **24** 189–95.
Dhar A, Jagadish C and Atherton D L (1992) *Mater. Eval.* **50** 1139–41.
Dodd C V (1977) *Research Techniques in Nondestructive Testing*, vol 3, ed R S Sharpe (London: Academic) Ch 13.
Dodd C V, Cheng C C and Deeds W E (1974) *J. Appl Phys.* **45** 638–47.
Dodd C V and Deeds W E (1968) *J. Appl. Phys.* **39** 2829–38.
Dodd C V, Deeds W E and Chitwood L D (1988) *Mater. Eval.* **46** 1592–97.
Dover W D and Collins R (1980) *Br. J. NDT* **22** 291–95.
Dubois J M S, Atherton D L, Czura W and Schmidt T R (1992) *BJNDT* **34** 401–6.
Duffin W J (1980) *Electricity and Magnetism*, 3rd edn (London: McGraw-Hill).
Edge B (1993) *Radar–Principles, Technology, Applications* (Englewood Cliffs, NJ: Prentice Hall).
Emerson P J (1976) *Br. J. NDT* **18** 48–51.
Fitzpatrick G L, Thome D K, Shih E Y C and Shih W C L (1993) *Mater. Eval.* **51** 1402–7.

Fitzpatrick G L, Thome D K, Skaugset R L and Shin W C C (1996) *Rev. Prog. QNDE* **15** 1159–66.
Förster F (1981) *Materialprüf* **23** 372–78.
Förster F (1982) *Defektoskopiya* **11** 3–25.
Förster F (1983) *Mater. Eval.* **41** 1477–78.
Förster F (1986) *Nondestructive Testing Handbook*, vol 4, 2nd edn, ed R C McMaster (Columbus, OH: American Society for Nondestructive Testing) sections 4 and 5.
Förster F and Libby H L (1986) *Nondestructive Testing Handbook*, vol 4, 2nd edn, ed R C McMaster (Columbus, OH: American Society for Nondestructive Testing) section 6.
Frise P R and Sahney R (1996) *Insight* **38** 96–101.
Geller L B, Poffenroth D, Udd J E and Hutchinson D (1992) *Mater. Eval.* **50** 56–63.
Gibbs M and Campbell J (1991) *Mater. Eval.* **49** 51–59.
Gowers K R and Millard S G (1991) *BJNDT* **33** 551–56.
Grimberg R, Savin A, Leitoiu S and Craus M L (1996) *Insight* **38** 650–52 and 655.
Gros X E (1996) *Insight* **38** 492–95.
Gros X E and Lowden D W (1995) *Insight* **37** 290–93.
Groves D and Connell D (1985) *NDT Int.* **18** 85–88.
Gu J and Yu L Y (1990) *NDT Int.* **23** 161–64.
Guehring W H (1982) *Br. J. NDT* **24** 251–58.
Guettinger T W, Grotz K and Wezel H (1993) *Mater. Eval.* **51** 444–61.
Hague B (1934) *Alternating Current Bridge Circuits* (London: Pitman).
Hanasaki K and Tsukada K (1995) *NDT & E Int.* **28** 9–14.
Hands B A (1985) *Cryogenic Engineering* (London: Academic) pp 27–28.
Harnwell G P (1938) *Principles of Electricity and Magnetism* (New York: McGraw-Hill).
Hatlo J (1979) *Br. J. NDT* **21** 317–19.
Hajian N T and Blitz J (1986) *NDT Int.* **19** 333–39.
Hajian N T, Blitz J and Hall R B (1983) *NDT Int.* **16** 3–8.
Heck C (1974) *Magnetic Materials and their Applications* (London: Butterworth).
Hill R, Geng R S, Cowking A and Mackersie, J W (1991) *NDT & E Int.* **24** 179–86.
Hill R, Geng R S and Cowking A (1993) *BJNDT* **35** 225–31.
Huang Z J, Meng R L, Qui X D, Sun Y Y, Kulik J, Xue Y Y and Chu C W (1993) *Physica C* **217** 1–5.
Ida N (1986a) *Nondestructive Testing Handbook*, vol 4, 2nd edn, ed R C McMaster (Columbus, OH: American Society for Nondestructive Testing) section 3.
Ida N (1986b) *Nondestructive Testing Handbook*, vol 4, 2nd edn, ed R C McMaster (Columbus, OH: American Society for Nondestructive Testing) section 19.
Ida N (1995) *Numerical Modelling for Electromagnetic Non-Destructive Evaluation* (London: Chapman & Hall).
Jansen H J M and Festen M M (1995) *Insight* **37** 421–25.
Jansen H J M, van de Camp P B J and Geerdink M (1994) *Insight* **36** 672–78.
Jenkins F A and White H E (1976) *Fundamentals of Optics*, 4th edn (Tokyo: McGraw-Hill).
Jiles D C (1988) *NDT Int.* **21** 311–19.
Johnston D (1992) *Mater. Perform.* **31** 36–39.
Junker W R and Mott G (1988) *Mater. Eval.* **46** 1353–57.
Kaneko T, Yamauchi H and Tankara S (1991) *Physica C* **178** 377–82.
Kaup P G and Santosa F (1995) *J. NDE* **14** 127–36.
Kaye G W C and Laby T H (1973) *Tables of Physical and Chemical Constants*, 14th edn (London: Longman Green).
Kirk I and Lewcock A (1995) *Insight* **37** 17–20 and 24.

Konev V A, Lyubetsky N V and Tikhanovich S A (1989) *Proc. 12th World Conf. on Non-Destructive Testing* (Amsterdam, 1989) vol 2, eds J Boogaard and G M van Dijk (Amsterdam: Elsevier) pp 1630–32.
Krause T W and Atherton D L (1994) *NDT & E Int.* **27** 201–7.
Krautkrämer J and H (1983) *Ultrasonic Testing*, 3rd edn (Berlin: Springer).
Krzwosz K R, Beissner R E and Doherty J E (1985) *Electromagnetic Methods of Nondestructive Testing*, ed W Lord (New York: Gordon and Breach) pp 307–20.
Kwun H and Holt A E (1995) *NDT and E Int.* **28** 211–14.
Lange L and Mook G (1994) *NDT & E Int.* **27** 241–47.
Langman R (1981) *NDT Int.* **14** 255–62.
Lawn B R and Wilshaw T R (1975) *Fracture of Brittle Solids* (Cambridge: Cambridge University Press).
Lebrun B, Jayet Y and Baboux J C (1995) *Mater. Eval.* **53** 1296–1300.
Lees D L, Pinder L W and Robins R H (1978) *Br. J. NDT* **20** 191–98.
Libby H L (1971) *Introduction to Electromagnetic Nondestructive Test Methods* (New York: Wiley–Interscience) Ch 8.
Lord W and Hwang J H (1977) *Br. J. NDT* **19** 14–18.
Lord W and Palanisamy R (1981) *Eddy Current Characteristics of Materials and Structure*, eds G Birnbaum and G Free (Philadelphia, PA: American Society for Testing and Materials) pp 5–21.
Lovejoy D J (1993) *Magnetic Particle Inspection* (London: Chapman & Hall).
Lovejoy D J (1994) *BJNDT* **36** 8–9.
Lugg M and Raine A (1995) *Insight* **37** 436–39.
Mackintosh D D, Atherton D L, Schmidt T R and Russell D E (1996) *Mater. Eval.* **54** 652–57.
McLachlan N W (1934) *Bessel Functions for Engineers* (London: Oxford University Press).
McMaster R C (ed) (1963a) *Nondestructive Testing Handbook*, vol 2 (New York: Ronald) section 35.
McMaster R C (ed) (1963b) *Nondestructive Testing Handbook*, vol 2 (New York: Ronald) section 54.
McMaster R C (ed) (1963c) *Nondestructive Testing Handbook*, vol 2 (New York: Ronald) sections 27 and 28.
McMaster R C (ed) (1963d) *Nondestructive Testing Handbook*, vol 2 (New York: Ronald) section 40.
McMaster R C (ed) (1963e) *Nondestructive Testing Handbook*, vol 2 (New York: Ronald) section 42.
McMaster R C (ed) (1963f) *Nondestructive Testing Handbook*, vol 1 (New York: Ronald) section 17.
Millard S G, Harrison J A and Edwards A J (1989) *Br. J. NDT* **31** 617–21.
Mills G, Priestley C, Reid R and Thirwell T (1983) *Br. J. NDT* **25** 127–29.
Moake G L and Stanley R K (1985) *Electromagnetic Methods of Nondestructive Testing*, ed W Lord (New York: Gordon and Breach) pp 151–60.
Molyneaux T C K, Millard S G, Bungey J H and Zhou J Q (1995) *NDT & E Int.* **28** 281–88.
Nash D C (1977) *Br. J. NDT* **19** 131–32.
Neumaier P (1983) *Br. J. NDT* **16** 233–37.
Oaten S R (1989) Assessment of defects in ferromagnetic materials with eddy currents, PhD Thesis, Brunel University, London.
Oaten S R and Blitz J (1987) *Nondestruct. Testing Commun.* **3** 139–51.
Owston C N (1985) *Br. J. NDT* **27** 227–31.
Padaratz I J and Forde M C (1995) *Nondestruct. Test. Eval.* **12** 9–32.

# References

Page L and Adams N I (1969) *Principles of Electricity*, 4th edn (Princeton, NJ: Van Nostrand).
Pautz J F and Abend K (1996) *Mater. Eval.* **54** 1004–8.
Perry C C and Lissner H R (1962) *Strain Gage Primer*, 2nd edn (New York: McGraw-Hill).
Pizzi P and Walther H (1979) *Br. J. NDT* **21** 16–22.
Poffenroth D N (1985) *Electromagnetic Methods of Nondestructive Testing*, ed W Lord (New York: Gordon and Breach) pp 35–69.
Rao S S (1989) *The Finite Element Method in Engineering*, 2nd edn (Oxford: Pergamon).
Roberts N, Nesbitt G and Fens T (1994) *Mater. Eval.* **11** 273–91.
Robertson I M (1994) *Acta Metall. Mater.* **42** 661–65.
Rollwitz W L (1973) *Nondestructive Testing: A Survey* NASA SP-5113 (Washington, DC: NASA) Ch 7.
Saadatmanesh H and Ehsani R (1997) *Insight* **39** 75–82.
Saadatmanesh H, Ehsani R and Yanez J C (1995) *NDT & E* **12** 133–53.
Sather A (1981) *Eddy Current Characterisations of Materials and Structures* ASTM STP722, eds G Birnbaum and G Free (Philadelphia, PA: American Society for Testing and Materials) pp 374–86.
Scanlon M J B (ed) (1987) *Modern RADAR Techniques* (London: Collins). Suitable for general purposes.
Schmidt T R (1984) *Mater. Eval.* **42** 225–30.
Schmidt T R (1986) *Nondestructive Testing Handbook*, vol 4, 2nd edn, ed R C McMaster (Columbus, OH: American Society for Nondestructive Testing) section 8, part 1.
Schmidt T R (1989) *Mater. Eval.* **47** 14–22.
Schroeder G (1984) *Computers in Nondestructive Testing*, Report 243 (Reutlingen: Institut Dr Förster).
Silk M G (1977) *Research Techniques in Nondestructive Testing*, vol 3, ed R S Sharpe (London: Academic) Ch 2.
Simms S (1993) *Mater. Eval.* **51** 529–34.
Sipahi L B, Jiles D C and Chandler A (1993) *J. Appl. Phys.* **73** 5623–25.
Smith J H and Dodd C V (1975) *Mater. Eval.* **33** 279–83 and 292.
Smith M E (1994) *Mater. Eval.* **11** 249–54.
Stanley R K (1986) *Nondestructive Testing Handbook*, vol 4, 2nd edn, ed R C McMaster (Columbus, OH: American Society for Nondestructive Testing) section 21.
Stanley R K (1996) *Insight*, **38** 51–55.
Stanley R K, Hiroshima T and Mester M (1986) *Nondestructive Testing Handbook*, vol 4, 2nd edn, ed R C McMaster (Columbus, OH: American Society for Nondestructive Testing) section 22.
Strange J H (1994) *Mater. Eval.* **11** 261–71.
Stumm W (undated) (Reutlingen: Institut Dr Förster) Report 226.
Stumm W (1985) *Electromagnetic Methods of Nondestructive Testing*, ed W Lord (New York: Gordon and Breach) pp 330–36.
Su W, Hazim O A, Al-Quadi A L and Riad S M (1994) *Mater. Eval.* **52** 496–502.
Suhr H and Guettinger T W (1993) *BJNDT* **35** 634–38.
Sullivan S, Atherton D L and Schmidt T R (1990) *Br. J. NDT* **32** 71–75.
Sun Y S, Udpa S, Lord W and Cooley D (1996) *Mater. Eval.* **54** 510–12.
Taylor J L (ed) (1988) *Basic Metallurgy for Non-Destructive Testing*, revised edn (Northampton: British Institute of Non-Destructive Testing).
Tilson J P and Blitz J (1985) *Br. J. NDT* **27** 276–78.
Topp D A (1994) *Insight* **36** 422–25.
Valdes L B (1954) *Proc. IRE (New York)* **42** 420–27.

Voskreseneskii D I and Voronin E N (1987) *Sov. Phys. Dokl.* **32** 411–12.
Waidelich D L (1981) *Eddy Current Characterisations of Materials and Structures* STM STP722, eds G Birnbaum and G Free (Philadelphia, PA: American Society for Testing and Materials) pp 367–73.
Weischedel H R and Chaplin C R (1991) *Mater. Eval.* **49** 362–67.
Wincheski B, Fulton J P, Nath S, Namkung M and Simpson J W (1994) *Mater. Eval.* **52** 22–26.
Wittig G and Thomas H-M (1981) *Eddy Current Characterisations of Materials and Structures* ASTM STP722, eds G Birnbaum and G Free (Philadelphia, PA: American Society for Testing and Materials) pp 387–97.
Woods R C (1996) *Eng. Sci. Educ. J.* **5** 51–56.
Wright D A (1966) *Semi-Conductors*, 4th edn (London: Science Paperbacks & Methuen).
Wu M K, Ashburn J R, Torng C J, Hor P H, Meng R L, Gao L, Huang Z J, Wang Y Q and Chu C W (1987) *Phys. Rev. Lett.* **58** 908–10.
Zick E (1994) *Mater. Eval.* **11** 255–60.
Zoughi R (1995) *Mater. Eval.* **53** 461–462.

# Index

AC bridge 32–5, 133–4
AC field method (ACFM) 217–23
  crack microgauge 219–21
  probe 220–1
AC potential drop method (ACPD) 217–23
AC power 29–30
ACFM crack microgauge 220–2
Alternating currents 23–31
Anisotropy measurements 209, 212
Attenuation coefficient (distance) 41, 42
Attenuation coefficient (time) 26
Automated NDT 9–10, 162–3
  pipelines 78

Barkhausen effect 37–8, 83–5
Bessel functions 237–40
Boundary frequency 101
Brewster angle 189–205
Bridges
  capacitance 35
  inductance 33, 133–4
  Kelvin double 227–8
  resonance 136–7
  Wheatstone 33, 227

Capacitance
  bridge 35
  measurements 282–231
Capacitors 23–5
CeNteSt 85–6
Circoflux 74–8
Circuit networks 21–2
Coating thickness measurements 146, 149–50, 153–5, 207
Coercivity (coercive force) 37, 85
Coils, eddy current *see* Eddy current testing, probes
Composition of materials 6, 13, 82, 84–5, 227–231

Computer prediction of impedance components
  coil encircling conducting rod 241–2
  coil encircling conducting tube 243–7
  air-cored surface-scanning coil 248–9
Concrete testing 222
Condition monitoring 3
Conductivity, electrical *see* Electrical conductivity
Corrosion 85–6, 153, 167–70, 217, 222
Coulomb's law 15, 20
Crack detection 136–7
Cracks, inclined 121–2
Curl operator 21–2

Debris testing 91
Decibel 30
Defect modelling (or simulation) 157–61
  eddy current testing 121–31
    mechanical 124–7
    numerical 128–31
Defects
  cracks 136–7, 214–20, 215–7
  distinguishing inner and outer surface defects 75–8
  internal defects 7–9
  lamina (shallow) 142–3
  rods and tubes 60–1, 74–7, 157–60, 163–6
  surface defects 62–3, 78–9, 225–7
Demagnetization 37, 45–6, 61–2
Dielectrics 15–19
  dielectric constant 18–19
  dielectric displacement 16–18, 23
  dielectric loss 19, 41
  testing of dielectrics 185–213, 225–231
Dimension measurements 7, 88–91, 101–3, 144–6, 152–5, 184–90, 199, 202–7, 215–20, 230
Dye penetrant testing 8–9

Eddy current testing 94–183
  advanced, see Eddy current testing, more advanced
  advantages 94
  artificial defects 94–183
  automatic 162–163
  basic tests 132–4, 149–61
  calibration 95, 134–6, 157–160, 164–6
  choice of equipment 144
  choice of frequency 147–9
  coils, see Eddy current testing, Probes
  defect imaging, see Eddy current testing, Magneto-optic eddy current imaging (MOI)
  defect modelling 121–31
    mechanical 124–7
    mercury 124–6
    Wood's metal 126–7
    numerical 128–31
  fastener holes, testing 162
  fibre-reinforced plastics 177–9
  fundamental measurements 132–4
  hand-held instrument 154–7
  impedance analysis 96–121
    computer programs 241–50
    encircling coils 96–108
      rods 96–104
      tubes 104–8
    internal axial coils 108–9
    surface scanning coils 109–21
  lift-off effects, elimination 143
  lift-off flaw detection 171–3
  magnetic saturation 102, 143
  magneto-optic eddy current imaging (MOI) 181–3
  methods 132–183
  microwave 176
  more advanced 162–183
  multifrequency 163–7
  neural networks 179–180
  normalization
    advantages 101
    frequency 101, 110–1, 123, 150
    impedance 98, 105, 106–7, 109, 116–20
    lift-off 110–1, 118–20
  principles 94–131
  probes 94–5, 109–11, 137–47, 157–8
    gap 143
    reflection 146–7
    self-nulling 149–51
    shielded 141, 151
    transformer 138
  pulsed 173–176
  remote field testing 167–170
  requirements 144–149
  standards 144
  surface scanning tests 160–1
  test pieces 157–60
  through-transmission method 138, 150, 151
  tube and rod testing 157–9, 163–70

Electric field 15–16
Electric flux 16–19
Electric flux density 16–18, 23
Electrical conductivity
  measurements 101–2, 118–20, 149–51, 215–7
Electrical permittivity
  measurements 207, 228–31
Electrical resistivity 13–15
Electric/magnetic field interactions 21–3
Electrified particle testing 225–227
Electromagnetic microwave radiation, see Microwave radiation
Electromagnetic penetration depth 43
Electromagnetic radiation 39–43, see also Microwave radiation
Electromagnetism 19, 25

Faraday effect 181–3
Faraday's law 21
Fastener holes, testing 162
Ferrites 35
Ferrography 91
Ferromagnetic materials 35–9
Ferromagnetic resonance probe 198–9
Fibre reinforced plastics 177–9
Fill factor 99, 107, 109
Finite element analysis 128–131
Flux leakage, see Magnetic flux leakage testing
Frequency
  boundary 101
  limiting 101
  normalized 101, 110–1, 123, 148
Frequency response 28–9

Gauss's theorem 18–21
Gunn diode 197–8

Hall effect 67–9
Hall probe 67–9
Hardness, see Structure of materials
Heat treatment, see Structure of materials
Hysteresis, see Magnetic hysteresis

Impedance, AC 28
  matching 133–4

# Index

Impedance component predictions for eddy current coils using computer programs
  coil encircling a cylindrical rod 241–2
  coil encircling a cylindrical tube 243–7
  surface scanning coil 248–50
Incremental permeability method 85–7
Inductance
  mutual 30–1
  self 21
Inductance bridge 34, 133–4
Intelligent pig 78–80

Josephson effect 72
Josephson Junction 72

Kelvin double bridge 227–8
Kirchhoff's law 31–2
Klystron 197
Koerzimat 85

Lift-off 110–1, 118–20
Limiting frequency 101

Magnatest 87–9
Magnetic balance 89–90
Magnetic circuits 38, 39, 46
Magnetic coercivity (coercive force) 36
Magnetic debris testing 91
Magnetic/electric field interactions 20–3
Magnetic field 19–21
Magnetic flux 20
Magnetic flux leakage testing 44–81
  see also Magnetic particle inspection (MPI)
  magnetic saturation 75
  principles 44–8
  quantitative methods 67–81
    detectors 67–74
    eddy current detector 74
    SQUID magnetometer 71–4
    standards 234
Magnetic hysteresis 35–8, 82–9
  Barkhausen effect 37–8, 83–5
  B/H loop 82–3
  coercivity (coercive force) 85
  incremental permeability 85–7
  Magnatest 87–9
  testing 82–9
Magnetic methods 43–93
Magnetic moment 46
Magnetic particle inspection (MPI) 48–65
  advantages and disadvantages 48–53
  automatic inspection 63
  demagnetization 61–2
  field induction methods 60–1
  inks 52–3
  magnetic field excitation 53–61
    current flow methods 57–61
    electromagnets 57–61
    field induction methods 60–1
    permanent magnets 53–7
    residual field method 54
    threading bars and coils 60–1
    yoke methods 54–7
  magnetic putty 64–5
  methods of detection 52–3
  multiple tests 53
  sensitivity of detection 53
  standards 234
  testing procedures 62–3
  underwater applications 64–5
Magnetic permeability 20, 35–8
  effective relative 98–100, 106, 109
  incremental 85–7
  measurements 120, 151–2
  recoil 103
  incremental permeability method 85–6
Magnetic reluctance 38, 90–1
  measurement 90–1
Magnetic retentivity (remanence) 36
Magnetic saturation 36
  eddy current testing 103, 143
  magnetic flux leakage testing 75
Magnetic tape inspection 65–7
Magnetic vector potential 23, 111
Magnetization 35–8
Magnetizing methods 82–7
Magnetodiode 69–70
Magneto-optic eddy current imaging (MOI) 181–3
Magnetoresistance 69–70
Magnetoresistive element 71
Magnetostriction 91–2
Maxwell's inductance bridge 133–4
Materials, composition, see Composition of materials
Microprobe 70
Microwave eddy current testing 176
Microwave methods, see Microwave radiation and Microwave testing
Microwave radiation 186–195
  absorption 194–5
  attenuation 194–5
  cavity resonator 191–2

Microwave radiation (contd)
    reflection and transmission at boundary
        single boundary
            normal incidence 187–9
            oblique incidence 189
        two parallel boundaries
            normal incidence 192–4
    relaxation 195
    resonance 189–192
    scattering 195, 206
    stationary waves 189–192
    see also Microwave testing
Microwave testing 186–195
    alternative to ultrasonic testing 185–6, 211
    applications 185–6
    attenuation measurements 207–8
    circuits 201–2
    detectors 198–9
    eddy current microwave testing 176
    fixed frequency continuous progressive wave method 203–6
    fixed frequency continuous stationary wave method 203–6
    highly attenuating materials 204–5
    horns 200–201
        near and far fields 200–1
    impedance 200
    instrumentation 195–202
    polarimetry measurements 212
    progressive wave methods 203–6
        oblique incidence 205–6
    RADAR 212–13
    refractive index measurements 207
    resonance cavity method 208–9
    safety 184–5
    scattering measurements 206
    sources 197–9
    stationary wave methods 206–8
    swept frequency echo method 210–11
    thickness measurements 207
    variable frequency methods 208–11
    waveguides 199–200
    see also Microwave radiation
Moisture content 195, 200

NDT methods 4–5
    choice of 5–9
Networks, circuit 31–2
Neural networks 179–180
Non-destructive evaluation (NDE) 3
Non-destructive testing (NDT) 1
    methods 4–5
        choice of 5–9

Normalization, see Eddy current testing, normalization
Nuclear magnetic resonance 92–3

Oscillatory circuits 25–8

Penetration depth, electromagnetic 43, 110
Permeability, see Magnetic permeability
Permittivity, see Electrical permittivity
Pig, intelligent see Intelligent pig
Pipeline inspection 78–81, 85–7, 167
Plate testing 80–1
Potential difference 15–6
Potential drop methods 214–223
    AC 217–223
    DC 215–17
    see also AC field method
Power, AC 29–30
Probe, Hall, see Hall probe
Probes, eddy current, see Eddy current testing, probes
Pulsed eddy currents, see Eddy current testing, pulsed

Quality control 1

RADAR testing 212–13
Radiological methods 6–7
Reflex klystron 197
Reluctance, magnetic 38
Remote field eddy current testing, see Eddy current testing, remote field testing
Residual magnetism 48
Residual stress 227
Resistance measurements 227–8
Resistance strain gauge 223–5
Resistivity, electrical, 13–15
    CNS resistivity meter 222–3
Resonance
    electric circuit 28–9
    microwave 191–2
Resonance bridge 136–7
Rod testing 74–8, 157–8
Rotomat 74–7

Sensors, magnetostrictive 69–70
Sigmaflux 133
Skin depth 42
SQUID 71–4
    DC 72–3
    RF 72–4
Standards
    eddy current 235–6
    flux leakage 234
    magnetic particle inspection 234

# Index

Statiflux 226
Strain gauge, resistance 223–5
Superconduction 71–4
  low temperature 71
  high temperature 72
Steel-wire rope inspection 81
Stress measurement 91–2
Stress, residual 227
Structure of materials 5–7, 82–9, 149, 228–31

Testing, automatic, *see* Automated NDT
Thickness measurement 146–7
  *see also* Coating thickness measurements

Transomat 75
Tube testing 74–81, 85–7, 157–9, 162–170
  resolving internal/external defects 74–5, 157–61

Ultrasonic methods 6–9
Underwater testing 220

Vector potential 111

Wheatstone bridge 33, 227
Wire testing 81